A LONG WAY FROM EUCLID

By the Author

From Zero to Infinity
Introduction to Higher Mathematics
A Long Way from Euclid

1·28·82

A
LONG
WAY
FROM
EUCLID

 by Constance Reid

Thomas Y. Crowell Company   New York

 Established 1834

Designed by Judith Woracek Barry

Manufactured in the United States of America

Library of Congress Catalog Card No. 63-18418

5 6 7 8 9 10

This book, which is based in part on the author's *Introduction to Higher Mathematics*, has been written for those whose first, and last, contact with real mathematics was with plane geometry and the *Elements* of Euclid. In a sense mathematics as we know it today began with the *Elements*. In more than two thousand years it has, of course, come a long way from Euclid. But it has never left him behind.

It is the hope of the author that the reader of this book will be able to glimpse through his own misty memories of Euclid's geometry the outline of some of the more imposing edifices of modern mathematics.

CONTENTS

vii

viii

There is no royal road to geometry.
—EUCLID TO PTOLEMY I

Modern geometry is a royal road.

1

*The
Golden Knot
in the
Golden Thread*

IN ANCIENT GREECE, WHERE MODERN mathematics began, there was no question among mathematicians but that the gods themselves were mathematicians too. But were the gods arithmeticians, or were they geometers?

Number ruled the Universe, according to Pythagoras in 500 B.C. Two centuries after Pythagoras, at about the same time that Euclid was compiling the *Elements,* Plato was asked, "What does God do?" and had to reply, "God eternally geometrizes." The choice of God as geometrician rather than arithmetician had quite literally been forced upon Plato and the other Greeks by two of the profoundest achievements of pre-Euclidean mathematics, both of them—ironically—due to Pythagoras and his followers.

These two achievements determined the decisive choice of form over number and set Western mathematics on the path it would follow for twenty centuries. The first was the discovery—and proof—that the square on the hypotenuse of a right triangle is equal to the sum of the squares on the other two sides. The second was the discovery—and proof—that when the sides of a right triangle are equal there is no number which exactly measures the length of the hypotenuse.

1

Specific instances of what we now call the Pythagorean theorem were known long before the Greeks in such far and separated parts of the world as India and China, Babylon and Egypt. In early Egypt, as the pyramids were being erected, basic right triangles were formed on the knowledge of the most familiar instance of the theorem:

$$3^2 + 4^2 = 5^2$$

A rope was divided into twelve units by knots tied at equal intervals, and pegs were placed in the third, seventh, and final knots. When the rope was stretched and pegged into place, it formed of necessity the desired right triangle:

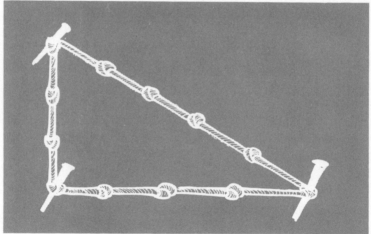

Although the Egyptians knew $3^2 + 4^2 = 5^2$ and other similar relationships obtained by multiplying or dividing this one, we do not know if they were aware that the equation gave no mere approximation but a theoretically exact right triangle.*

Whether this general truth was actually known

* The Rhind Papyrus contains only such equivalents as $(\frac{3}{4})^2 + 1^2 = (1\frac{1}{4})^2$ and $12^2 + 16^2 = 20^2$, which are obtained, respectively, by dividing and multiplying the original by 4.

2

earlier, history has left the discovery of the general theorem to the Greeks, and traditionally to Pythagoras. Pythagoras was in his youth a pupil of Thales, who had measured the height of the great pyramid by comparing the length of its shadow with that of a vertical stick. Later, as a teacher himself, Pythagoras opened a school of his own in his native town, where he attracted only one pupil, also named Pythagoras, whom he had to pay to keep in class. Justifiably discouraged by this lack of appreciation at home, he set out, as Thales had once advised him, for Egypt. He came at last, after years of travel and study, to southern Italy. Here he opened a school which, in contrast to his first, was one of the most wildly successful schools in history. Crowds flocked to hear Pythagoras. Besides the youths whom he instructed during the day, the business and professional leaders of the community attended his evening lectures and—to hear Pythagoras—maiden and matron alike broke the law which prohibited them from attending public meetings.*

The teachings of Pythagoras were something of a mixture—almost equal parts of morality, mysticism and mathematics. He saw life as a precarious balance of ten somewhat random but nevertheless fundamental pairs of opposites: odd and even, limited and unlimited, one and many, right and left, male and female, rest and motion, straight and curved, light and darkness, good and evil, square and oblong. It was a particularly happy circumstance for Pythagoras that the number of these fundamental opposites was 10, for from his point of view 10 was the most perfect of numbers, being the sum of 1 (the point), 2 (the line), 3 (the plane) and 4 (the solid).

Pythagoras and his followers were people who saw Number in every relationship and very personal attributes

* One of these pupils, a young and beautiful (and intelligent) girl, married the sixty-year-old teacher.

3

in the individual numbers.* Their great discovery of the dependence of the musical intervals on certain arithmetic ratios of strings at the same tension provided scientific support for what they had always intuitively considered to be true:

Number rules the Universe.

To such a people even their everyday surroundings spoke of Number. Quite probably, the first general recognition of a particular instance of the famous theorem about the square on the hypotenuse occurred when someone saw this truth as it was exhibited in the regular checkered tiling of a floor. From inspection it would have been clear that the square on the diagonal of any tile contained as many half-tiles as the squares on both sides put together:

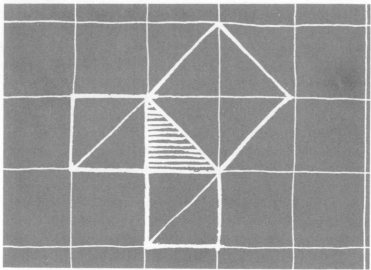

It would also have been clear that this relationship be-

* The number 1 stood for reason; 2, for opinion. There is no record that 3 represented disagreement; but 4, at any rate, was the number of justice.

4

tween the diagonal (or the hypotenuse of the right angle) and the sides would remain true regardless of the size of the individual squares.

A square cut by a diagonal represents only one particular kind of right triangle—that in which the two sides containing the right angle are equal. But no one who is at all mathematically inclined, today or twenty-five hundred years ago, could observe such a truth about isosceles right triangles without wondering if it applied as well to *all* right triangles. Thus the general theorem would be suggested:

THEOREM: *The square on the side of the hypotenuse of a right triangle is equal to the sum of the squares on the other two sides.*

To make such a statement about right triangles, either we must verify it by actually examining all right triangles (which is impossible, since there are an infinite number of them) or we must prove that it is a necessary consequence of right triangle-ness and, therefore, has to be true of all right triangles.

In the centuries since the discovery of this theorem, there have been literally hundreds of proofs of the fact that the square on the hypotenuse of any right triangle is equal to the sum of the squares on the other two sides.*
At one time, a completely new proof was a requirement for a master's degree in mathematics.

No one knows exactly how Pythagoras himself proved the general theorem. The proof which appeared a few hundred years later in the *Elements* is definitely not

---

* A Mason who saw in the Pythagorean society the beginnings of Masonry made a classified collection of more than two hundred proofs of the famous theorem (E. S. Loomis, *The Pythagorean Proposition*) and gave the publication rights to the Masters and Wardens Association of the 22nd Masonic District of the Most Worshipful Grand Lodge of Free and Accepted Masons of Ohio.

Pythagorean, being the only theorem in the book which tradition universally ascribes to Euclid himself.

It would be pleasant to think that Pythagoras first established this great truth with one of those ingenious arrangements which bring the idea to eye and mind in the instant of seeing. Such a proof would be given by the two equal squares below with sides $(a + b)$. These show without a word that

$$a^2 + b^2 = c^2$$

since both sides of the equation, when subtracted from the two original and equal squares, leave as remainders four right triangles, all of the same size.

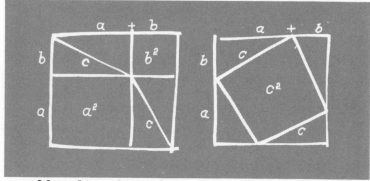

Although we do not know how the theorem was actually proved, tradition tells us that Pythagoras himself was so delighted (and certainly any true mathematician would have been!) that he sacrificed to the gods a hecatomb (100) of oxen, causing the theorem to be known during the Middle Ages as *inventum hecatomb dignum.**

* There is no specific written evidence that Pythagoras himself discovered or proved the theorem which bears his name. It was the custom for all discoveries of the school to be attributed to the master himself, regardless who made them. However, early writers are quite definite about "one famous figure" discovered by Pythagoras and "a famous proposition on the strength of which he offered

Thus, five hundred years before the birth of Christ, mathematics had in hand its famous theorem about the square on the hypotenuse of the right triangle—a theorem which was destined, in the words of E. T. Bell, to run "like a golden thread" through all of its history. This theorem would serve—in trigonometry, which is entirely based on it—as the tool for measurement lying beyond the immediate use of tape measure and ruler. In analytic geometry, it would serve as the basic distance formula for space in any number of dimensions. In its arithmetical generalization $(a^n + b^n = c^n)$, it would provide mathematics with its most famous *unsolved* problem, known as Fermat's Last Theorem.* In the most revolutionary mathematical discovery of the nineteenth century, it would be revealed as the equivalent of the distinguishing axiom of Euclidean geometry; and in our own century it would be further generalized so as to be appropriate to and include geometries other than that of Euclid. Twenty-five hundred years after its first general statement and

---

a splendid sacrifice of oxen." That the famous figure and the famous proposition were one and the same, and that both referred to the theorem about the square on the hypotenuse, is not certain. Tradition, however, has always insisted upon ascribing the theorem to the man Pythagoras.

* Toward the end of the sixteenth century, an "amateur" French mathematician named Pierre Fermat noted in the margin of a book of problems the theorem that $a^n + b^n = c^n$ is solvable if, and only if, $n = 2$ (i.e., as in the Pythagorean theorem). He did not prove his theorem but added regretfully to his note, "I have discovered a truly marvelous proof of this, which, however, the margin is not large enough to contain." Today it is generally thought that the theorem is true, but that Fermat was mistaken when he said he had discovered a proof. Efforts to prove Fermat's Last Theorem have resulted in the development of many extremely valuable mathematical methods; and it has been said that, if the margin of Fermat's book had been wider, the whole history of mathematics might have been different!

proof, the theorem of Pythagoras would be found, firmly embedded, in Einstein's theory of relativity.

But we are getting ahead of our story. For the moment we are concerned only with the fact that the discovery and proof of the Pythagorean theorem was directly responsible for setting the general direction of Western mathematics.

We have seen how the Pythagoreans lived and discovered their great theorem under the unchallenged assumption that *Number rules the Universe*. When they said *Number,* they meant whole number: 1, 2, 3, ... Although they were familiar with the sub-units which we call fractions, they did not consider these numbers as such. They managed to transform them into whole numbers by considering them, not as parts, but as *ratios* between two whole numbers. (This mental gymnastic has led to the name *rational numbers* for fractions and integers, which are fractions with a denominator equal to one.) Fractions disposed of as ratios, all was right with the world and Number (whole number) continued to rule the Universe. The gods were mathematicians—arithmeticians. But, all the time unsuspected, there was numerical anarchy afoot. That it should reveal itself to the Pythagoreans through their own most famous theorem is one of the great ironies of mathematical history. The golden thread began in a knot.

The Pythagoreans had proved by the laws of logic that the square on the hypotenuse of the right triangle is equal to the sum of the squares on the other two sides. They had also discovered the general method by which they could obtain solutions in whole numbers for all three sides of such a triangle. Although these whole number triples (the smallest being the long-known 3, 4, 5) still bear the name of "the Pythagorean numbers," the

Pythagoreans themselves knew that not all right triangles had whole-number sides. They assumed, however, that the sides and hypotenuse of any right triangle could always be measured in units and sub-units which could then be expressed as the ratio of whole numbers. For, after all, did not Number—whole number—rule the Universe?

Imagine then the Pythagoreans' dismay when one of their society, observing the simplest of right triangles, that which is formed by the diagonal of the unit square, came to the conclusion and proved it by the inexorable processes of reason, that there could be no whole number or ratio of whole numbers for the length of the hypotenuse of such a triangle:

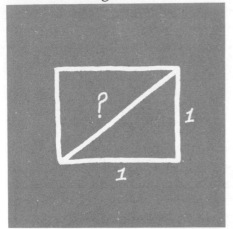

$$1^2 + 1^2 = 2$$
$$\sqrt{2} = ?$$

When we look at any isosceles right triangle—and remember that the size is unimportant, for the length of one of the equal sides can always be considered the unit of measure—it is clear that the hypotenuse cannot be measured by a whole number. We know by the theorem of Pythagoras that the hypotenuse must be equal to the square root of the sum of the squares of the other two sides. Since $1^2 + 1^2 = 2$, the hypotenuse must be equal to $\sqrt{2}$. Some number multiplied by itself must produce 2. What is this number?

It cannot be a whole number, since $1 \times 1 = 1$ and $2 \times 2 = 4$. It must then be a number *between* 1 and 2. The Pythagoreans had always assumed that it was a rational "number." When we consider that the rational numbers between 1 and 2 are so numerous that between any two of them we can always find an infinite number of other rational numbers, we cannot blame them for assuming unquestioningly that among such infinities upon infinities there must be some rational number which when multiplied by itself would produce 2. Some of them actually pursued $\sqrt{2}$ deep into the rational numbers, convinced that, somewhere among all those rational numbers, there must be one number—one ratio, whole number to whole number—which would satisfy the equation we would write today as

$$\left(\frac{a}{b}\right)^2 = 2$$

The closest they came to such a number was $^{17}\!/_{12}$, which when multiplied by itself produces $^{289}\!/_{144}$, or $2^{1}\!/_{144}$.

But one of the Pythagoreans, a man truly ahead of his time, stopped computing and considered instead another possibility. *Perhaps there is no such number.*

Merely considering such a possibility must be rated as an achievement. In some respects it was even a greater achievement than the discovery and proof of the famous theorem that produced the dilemma!

Perhaps there is no such number. How does a mathematician go about *proving* that there isn't a solution to the problem he is called upon to solve? The answer is classic. He simply assumes that what he believes to be

10

false is in actuality true. He then proceeds to show that such an assumption leads to a contradiction, usually with itself, and of necessity cannot be true. This method has been vividly called proof *per impossibile* or, more commonly, *reductio ad absurdum*. "It is," wrote a much more recent mathematician than the Pythagorean, "a far finer gambit than a chess gambit: a chess player may offer the sacrifice of a pawn or even a piece, but a mathematician offers *the game*." *

The most recent proof † to shake the foundations of mathematical thought was based on a *reductio* and so, twenty-five hundred years ago, was the first. We shall present this proof, which is a fittingly elegant one for so important an idea, in the notation of modern algebra, although this notation was not available to the man who first formulated the proof.

Let us assume that, although we have never been able to find it, there actually is a rational number a/b which when multiplied by itself produces 2. In other words, let us assume there exists an a/b such that

$$\frac{a}{b} \times \frac{a}{b} = 2$$

We shall assume (and this is the key point in the proof) that a and b have no common divisors. This is a perfectly legitimate assumption, since if a and b had a common divisor we could always reduce a/b to lowest terms. Now, saying that

$$\frac{a}{b} \times \frac{a}{b} = 2$$

* G. H. Hardy, *A Mathematician's Apology* (Cambridge, England: Cambridge University Press, 1941).
† This proof, by the twentieth-century mathematician Kurt Gödel, will be discussed in the last chapter.

is the same as saying that

$$\frac{a^2}{b^2} = 2$$

If we multiply both sides of this equation by b^2 (which we can, since b does not equal 0 and since we can do anything to an equation without changing its value as long as we do the same thing to both sides), we shall obtain:

$$\frac{a^2 b^2}{b^2} = 2b^2$$

or, by canceling out the common divisor b^2 on the left-hand side:

$$a^2 = 2b^2$$

It is obvious, since a^2 is divisible by 2, that a^2 must be an even number. Since odd numbers have odd squares, a also must be an even number. If a is even, there must be some other whole number c which when multiplied by 2 will produce a; for this is what we mean by a number being "even." In other words,

$$a = 2c$$

If we substitute $2c$ for a in the equation $a^2 = 2b^2$, which we obtained above, we find that

$$(2c)^2 = 2b^2$$

or

$$4c^2 = 2b^2$$

Dividing both sides of this equation by 2, we obtain

$$2c^2 = b^2$$

Therefore, b^2, like a^2 in our earlier equation, must also be an even number; and it follows that b, like a, must be even.

12

BUT (and here is the impossibility, the absurdity which clinches the proof) we began by assuming that a/b was reduced to lowest terms. If a and b are both even, they must—by the definition of evenness—have the common factor 2. Our assumption that there can be a rational number a/b which when multiplied by itself produces 2 must be false, for such an assumption leads us into a contradiction: we begin by assuming a rational number reduced to lowest terms and end by proving that the numerator and the denominator are both divisible by 2!

We can only imagine with what consternation this result was received by the other Pythagoreans. Mysticism and mathematics were met on a battleground from which there could be no retreat and no compromise.* If the Universe was indeed ruled by Number, there must be a rational number a/b equal to $\sqrt{2}$. But by impeccable mathematical proof one of their members had shown that there could be no such number!

The Pythagoreans had to recognize that the diagonal of so simple a figure as the unit square was incommensurable with the unit itself. It is no wonder that they called $\sqrt{2}$ *irrational!* It was not a rational number, and it was contrary to all they had believed rational, or reasonable. The worst of the matter was that $\sqrt{2}$ was not by any means the only irrational number. They went on to prove individually that the square roots of 3, 5, 6, 7, 8, 10, 11, 12, 13, 14, 15 and 17 were also irrational.† Although they worked out a very ingenious method of approximating

* "He is unworthy of the name of man who is ignorant of the fact that the diagonal of the square is incommensurable with the side."—Plato, quoted by Sophie Germain, *Mémoire sur les surfaces élastiques.*

† The general theorem states that the square root of any number which is not a perfect square is an irrational number. According to an even more general theorem, the mth root of any number which is not a perfect mth power is irrational.

13

such irrational values by means of ratios (detailed on pages 14-15), they had to face the fact that there was not just one, there were many (in fact, infinitely many) lengths for which they could find no accurate numerical representation in a Universe that was supposedly ruled by Number.

Tradition tells us that they tried to solve their dilemma by persuading the discoverer of the unpleasant truth about $\sqrt{2}$ to drown himself. But the truth cannot be drowned so easily; nor would any true mathematician, unconfused by mysticism, wish to drown it. The Pythagoreans and the mathematicians who followed them, from Euclid to Einstein, had to live and work with the irrational.

Here was the golden thread impossibly knotted at its very beginning!

It was at this point that the Pythagoreans, rather than struggling to unravel arithmetically what must have seemed to them a veritable Gordian knot, took the way out that a great soldier was to take in a similar situation. They cut right through the knot. If they could not represent $\sqrt{2}$ exactly by a number, they could represent it exactly by a line segment. For the diagonal of the unit square *is* $\sqrt{2}$.

With a choice of two mathematical roads before them, the Greeks, long before the time of Euclid, chose the geometric one; and

"That has made all the difference."

FOR THE READER

Today we customarily approximate the value $\sqrt{2}$ by extracting the square root of 2 to as many decimal places as we feel necessary for accuracy. In this way, from one side, we approach closer and closer to that single point, which is represented by the non-terminating and non-

14

repeating decimal 1.41421. . . . Using rational representations rather than decimals, the Pythagoreans worked out a method of approaching this same point *from both sides* with successively closer approximations.

They began a ladder with a pair of 1's and by the additions indicated below obtained the number pairs on the right:

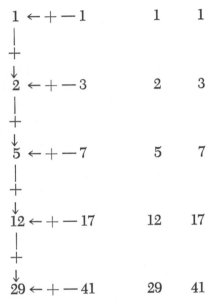

$1 \leftarrow + - 1$	1	1
$2 \leftarrow + - 3$	2	3
$5 \leftarrow + - 7$	5	7
$12 \leftarrow + - 17$	12	17
$29 \leftarrow + - 41$	29	41

The reader should try to determine the next rung of the ladder. If he will then square the fractions obtained by taking the numerator from the right and the denominator from the left, he will find that although he will never reach 2 exactly he will approach it in a continuously narrowing zigzag as the fractions he is squaring approach $\sqrt{2}$.

2

*Nothing,
Intricately
Drawn
Nowhere*

"A POINT IS THAT WHICH HAS NO PART."

Thus begins the most durable and influential textbook in the history of mathematics. Thus, in fact, begins modern mathematics.

It has been more than two thousand years since the Greek Eukleides, whom we know better as Euclid, gathered together the mathematical work of his predecessors into thirteen books which he entitled, simply, the *Elements*. During this time the *Elements* of Euclid, in addition to serving as a mathematical textbook for adolescents, has also served as Western man's final, and first, bulwark against ignorance. Newton cast his *Principia* in the already hallowed form of the *Elements*. Kant called on the axioms of the *Elements* as "the only immutable truths." On the first few pages of this seemingly spare and formal work, bloodless battles have been waged. It was here, at the middle of the nineteenth century, that mathematics made its greatest self-discovery; and it was here, at the beginning of the twentieth, that it made its great and final stand to establish—to prove, in fact—its own internal consistency. We have come, in the last two thousand years, a long way from Euclid; but we have also taken his *Elements* with us, all the way.

16

The man Euclid and the facts of his life and career were lost very early on the journey. We are told that he "flourished" about 300 B.C., that he founded a school at Alexandria in the time of Ptolemy I. There are about him only two traditional anecdotes, both of which are also recounted of other Greek mathematicians. In the years after his death various writers confused him with another Euclid, the philosopher of Megara; and the Arabs put forth a claim that he had really been an Arab all along. It can be said that in the history of mathematics there is no Euclid; there is only the *Elements*. Probably within his own time (in the words that Auden used of Yeats) he had become his admirers.

The *Elements*, from the beginning, was immediately recognized for what it was—a masterpiece. The form of the book was not original. The logical ladder of definitions, axioms, theorems and proofs was first erected by some earlier Greek than Euclid, perhaps a priest. The subject matter was not original. The masterly treatment of proportion which enabled the later Greeks to handle incommensurable as well as commensurable magnitudes, is that of Eudoxus; and the other books are frankly based on the known work of other men. ("The picture has been handed down of a genial man of learning, modest and scrupulously fair, always ready to acknowledge the original work of others," H. W. Turnbull wrote in *The Great Mathematicians*.) Only one proof—that of the Pythagorean theorem—is traditionally ascribed to Euclid himself, although it is apparent that to fit theorems into his new arrangement he must have had to create other new proofs. Even the title, the *Elements*, was not original. This term did not refer, as we might think, merely to the elementary aspects of the subject but rather—according to an early mathematical historian—to certain leading theorems in the whole of mathematics which bear to those which follow the relation of a principle, furnishing proofs of many properties. Such theorems were called by the name of elements; and their function was somewhat like that of the letters of the alphabet in the language, letters being called by the same name in Greek. There had been many *Elements* before Euclid. That there was none after him is an unequivocal tribute to the sheer genius of his work.

As a mathematician, Euclid falls far behind Eudoxus, who preceded him, and Archimedes and Apollonius, who followed. The *Encyclopaedia Britannica* admits regretfully that he was not even a "first-rate" mathematician, but adds that there is no question but that he was a first-rate teacher. What he brought to the already great mathematics of his time was a genius for system. And system was exactly what was needed! There were many fine single works on specialized subjects. Many editors had gathered together what seemed to them important. There were definitions, axioms, theorems and proofs galore; and an almost equal number of organized and disorganized, overly complete and incomplete arrangements, all called the *Elements*. Euclid took these. He selected, substituted, added, rearranged; and what came out in his *Elements* was a distillation of all those that had come before—a model of systematic thought.

We have no copy of this original work. Oddly enough, we have no copy made even within a century or two of Euclid's time. Until recently the earliest known version of the *Elements* was a revision with textual changes and some additions by Theon of Alexandria in the fourth century after Christ, a good six centuries after Euclid compiled it in Alexandria. Early in the nineteenth century, a Greek manuscript in the Vatican was discovered by

internal evidence to be a pre-Theonine text.* The traditional textbook version of the *Elements,* which was used
almost completely without change until very recently, was
based, of course, on the text of Theon. In a quite literal
sense, Euclid has become his admirers; for when we say,
"Euclid says," we are speaking of a compiler much closer
to us than the original compiler of the *Elements.* This is
unimportant at this time. We are not concerned with what
Euclid himself actually wrote in the *Elements,* but with
what has served mathematics for so many centuries as the
Elements of Euclid.

What, then, is this work which has played such an
influential role in the history of mathematics—and of
thought itself? Most of us are probably not familiar with
a translation of Theon's traditional version of the masterpiece. Our high school geometry textbook, however, was
probably based directly upon it. After a few introductory
remarks and simple explanations in modern terms most
authors in the past fell back very quickly upon the original. If we were to examine at this time a translation of
the *Elements,* such as Sir Thomas Heath's, available now
in paperback (Dover Press), we would find it unexpect-
edly familiar.

The *Elements*—we would find—is composed of thirteen sections, or "books," arranged according to subject
matter: the first few to plane geometry, the last to solid,
and the books between to proportion and number. We

* It is interesting to note that the Romans never translated the
Elements into Latin. "Among the Greeks," Cicero wrote con-
descendingly, "nothing was more glorious than mathematics. But
we have limited the usefulness of this art to measuring and calculat-
ing." The earliest extant Latin translation (c. 1120) is one by the
Englishman Athelhard, who obtained an Arabic copy of the *Elements* by going to Spain disguised as a Moslem student, and made
his translation from that copy.

would meet again the famous *pons asinorum*, or Bridge of Asses, as the fifth proposition in Book I:

THEOREM: *In isosceles triangles the angles at the base are equal to one another, and, if the equal straight lines be produced further, the angles under the base will be equal to one another.*

This is the theorem which traditionally separates mathematical boys from mathematical men, since the asses supposedly cannot get through the proof, or across the bridge. In the Middle Ages the mastering of this theorem and its proof marked the culmination of the mathematical training required for a degree.

At the end of Book I we would find our old friend, the famous theorem about the square on the hypotenuse of the right triangle, which laymen know as the theorem of Pythagoras and which loving geometers have called for over two thousand years merely "I, 47," because of its position as the forty-seventh proposition in the first book of Euclid's *Elements*. The proof of this theorem is the only one in the *Elements* which is specifically credited to Euclid himself. Although the philosopher Schopenhauer dismissed it contemptuously as a "mouse-trap proof" and "a proof walking on stilts, nay, a mean, underhand proof," Sir Thomas Heath, the English editor of the *Elements*, calls it "a veritable *tour de force* which compels admiration." It is Heath's contention that Euclid found the theorem proved by the incomplete theory of proportion of the Pythagoreans (incomplete because it was not applicable to the yet undiscovered incommensurable magnitudes), and that this proof by proportion suggested to him the method of I, 47. Although his plan for the *Elements* did not call for the treatment of proportion until Book V, according to the Heath theory, he managed to transform the Pythagorean proof by proportion into

20

one based on Book I only. "A proof extraordinarily ingenious," insists Heath—and a fig to the philosopher who expects an intuitive proof of the "look-see" type from the compiler of the *Elements!*

In Book V, we would find what is without question the finest mathematics in the *Elements*—the theory of proportion as expounded by Eudoxus. It was this theory, applying as it did to incommensurable as well as to commensurable magnitudes, which allowed Greek mathematicians, after the shattering discovery of the irrational, to move forward again. Because of its importance to our story as a whole, we shall treat it separately in Chapter 4.

After Book VI, which also deals with problems of proportion, we would find the three books on the theory of numbers. Although the "numbers" seems strangely unfamiliar, since they are all represented by straight lines "in continued proportion," we would find here many familiar truths of our own school arithmetic. Proposition 1 of Book VII, for instance, gives us the standard method —still known as "Euclid's algorithm" *—for finding the greatest common divisor of two numbers, although in the *Elements,* with its generally geometric approach, it is "the greatest common measure." As Proposition 20 of Book X, the third and final book on numbers, we would find that most important and interesting truth: that the number of primes is infinite; in Euclid's words, "Prime numbers are more than any assigned multitude of numbers." (Proved in the next chapter.)

At the end of the thirteenth and final book of the *Elements* we would meet again the five regular solids, those bodies with which the Platonists identified all creation. In their philosophy the cube represented the earth; the octahedron, the air; the tetrahedron, fire; the icosahe-

* Detailed at the end of this chapter.

21

dron, water; and the dodecahedron, the Universe itself. Good Platonists always maintained that Euclid organized the *Elements* solely for the purpose of presenting the construction of the perfect figures, but this is obviously not true. The *Elements* contains a great deal, including the three books on arithmetic, which contributes nothing to these final constructions.

As we continue our re-examination of the *Elements,* we would note a certain pattern in the arrangement. Each of the thirteen books begins with a list of definitions of the terms which will be needed in it; the first book is preceded as well by a group of more or less obvious statements, or axioms; and each of the thirteen books consists of a related series of theorems which are proved by appealing to the authority of previously stated theorems, axioms, and definitions, all of these derived logically by the accepted rules of reason.

This is the ladder by which the Greeks believed that man could ascend to truth—and they believed it to be the only ladder:

L		Proofs		L
O		Theorems		O
G		Axioms		G
I		Definitions		I
C				C

As Euclid is reputed to have told the first Ptolemy when asked if there were no other, easier way than that of the *Elements:* "There is no *royal* road to geometry." Today we call Euclid's ladder the axiomatic method, and we still find it the ladder by which man can ascend most surely to truth. If our concept of the truth we reach is somewhat different from that of the Greeks, that is a story for a later chapter; for the moment we must concentrate on examining the rungs of the ladder with the eyes and minds of the men who built it.

To the Greeks, the definitions given by Euclid at the beginning of each book of the *Elements* were not statements of existence but merely descriptions. Existence of that which was defined had to be established by constructions which met the specifications laid out in the definitions. In the words of Aristotle: "Thus, what is meant by triangle the geometer assumes, but that it exists he has to prove." Accordingly, in Book I Euclid begins by producing the equilateral triangle which he has described in Definition 20. In Proposition 11 he constructs a right angle (Definition 10) and in Proposition 46, a square (Definition 22). Until these figures are actually constructed on the authority of the axioms and previously proved theorems, they are never used in the *Elements*.

There are, however, certain terms defined at the beginning of Book I which Euclid never produces "from scratch." These are terms the existence of which is specifically implied by the postulates: the point, the straight line and the circle—in short, his "subject matter." These are the objects in terms of which all the others have been defined. Among the other definitions, Euclid does describe these objects, but just for the record:

A point is that which has no part.

A straight line is a line which lies evenly with the points of itself.

A circle is a plane figure contained by one line such that all the straight lines falling upon it from one point among those lying within the figure are equal to one another; and this point is called the center of the circle.

He clearly recognizes that he will never be able to produce a point, a straight line or a circle unless he *assumes* before he begins that he can produce them.

"Let the following be postulated," he announces at the beginning of Book I:

To draw a straight line from any point to any point.

To produce a finite straight line continuously in a straight line.

To describe a circle with any center and distance.

On the *arbitrarily assumed* ability to do these three things, the ladder rests. We can join any two points, extend any straight line, describe about any center a circle of any size—because we have agreed that we can. To those who may object that any point which we put on paper will have by the nature of the instrument with which we must make it some "part"; that for the same reason any line which we draw cannot lie evenly on all its partless points; that the points on the boundary of any circle cannot be all the same distance from the center—to all those who object, we have in the postulates our unanswerable answer: *we can because we have begun by agreeing that we can.*

"It is ignorance alone that could lead anyone to try to prove the axioms." *

But we must never forget that the choice of the assumptions on which we are to rest our ladder to truth is a purely arbitrary one. Just as in a game we could, by agreement of all the players, make different rules under which to play (making, of course, a different game of it), so Euclid could have chosen other axioms, as we shall see in a later chapter. It was his choice, more than anything else, which was indicative of his genius.

What constitutes a well-chosen set of axioms? Since

* Aristotle.

24

long before Euclid chose his, men have discussed this question, and they have always been pretty well agreed. There is one absolute requirement: consistency. The axioms that we have chosen must never lead us into a contradiction. Beyond this essential requirement there are others that are more of a practical or an esthetic nature. A well-chosen set of axioms should exhibit such virtues as simplicity, economy, sufficiency, and a certain indefinable "importance."

We could discuss more precisely the characteristics of these characteristics; but the reader can probably get a much quicker and much more vivid picture of the requirements if he imagines himself in the following game situations and considers, not what constitutes a well-chosen set of axioms, but rather what is wrong with the rules of the game which he is playing:

He finds the rules hard to play by because they list many exceptions. (*Not simple.*)

He finds that one of the rules is unnecessary since it is already stated, although in quite different words, by another rule. (*Not economical.*)

He finds that under the rules he cannot make a move which seems necessary if the game is to be really interesting. (*Not sufficient.*)

He finds that there is a rule which forbids a certain move which is permitted by another rule. (*Not consistent.*)

He finds that the game played according to the rules is so uninteresting that, even when he wins, he feels very little satisfaction. (*Not important.*)

If we substitute for "rules of the game," "set of axioms" and for "moves," "theorems," we see that the requirements are very much the same; and the axioms Euclid chose so well in Alexandria long before the birth of Christ have provided Western man for more than twenty centuries with a very good game indeed.

Before we leave the subject of the axioms, we should point out that Euclid distinguished between two types of assumptions, "common notions" and "postulates." The common notions include such statements as "the whole is greater than the part"; while one of the postulates states that "all right angles are equal." (All the common notions and postulates are listed on page 27, since we shall be referring to them again from time to time.) Probably no one has been able to say exactly what distinction Euclid himself made between the two; but if anyone is well qualified to make an educated guess, it is Sir Thomas Heath, a career civil servant in the British government, who will go down in history as the ultimate and complete editor of the *Elements*.

Heath writes on the two different types of axioms: "As regards the postulates we may imagine him [Euclid] saying, 'Besides the common notions there are a few other things which I must assume without proof, but which differ from the common notions in that they are not self-evident. The learner may or may not be disposed to agree with them; but he must accept them at the outset on the superior authority of his teacher, and must be left to convince himself of their truth in the course of the investigation which follows.' "

Having defined our terms and agreed upon them and to our axioms (common notions and postulates alike), we are now ready to climb, rung by rung, the ladder of mathematical truth, guided always by the accepted laws of logic. Each rung of this ladder is a proposition (which may be either a problem or a theorem) and its proof; and by the rules of the game each rung may utilize in its con-

26

struction only the rungs below. This means that the first proposition must depend for its proof only upon the axioms and definitions already given, but the second may utilize as well the now proved first proposition, and so on. By the time we arrive at the famous fifth, the *pons asinorum*, we find that to prove it we need Propositions 3 and 4, which we have already proved, as well as Postulates 1 and 2. This process continues. The proof of I, 47,

AXIOMS AND POSTULATES OF EUCLID *
AXIOMS

1. Things which are equal to the same thing are also equal to one another.

2. If equals be added to equals, the wholes are equal.

3. If equals be subtracted from equals, the remainders are equal.

4. Things which coincide with one another are equal to one another.

5. The whole is greater than the part.

POSTULATES

Let the following be postulated:

1. To draw a straight line from any point to any point.

2. To produce a finite straight line continuously in a straight line.

3. To describe a circle with any center and distance.

4. That all right angles are equal to one another.

5. That if a straight line falling on two straight lines makes the interior angles on the same side less than two right angles, the straight lines, if produced indefinitely, will meet on that side on which the angles are less than two right angles.

* This version is given by Sir Thomas Heath in *The Elements of Euclid.*

27

relies upon five previously proved propositions as well as on two of the common notions which we agreed to before we started. From the moment, on the first page of the *Elements,* when we placed our hand on the first rung ("A point is that which has no part"), we have been climbing.

His own time considered the *Elements* of Euclid as near to perfect as work of man could be. The succeeding centuries of the Christian era were, as we shall see in a later chapter, troubled by one small flaw which they struggled valiantly to eliminate, only to find in the end that it supported the entire edifice (something Euclid himself had apparently known when he laid the foundations). At the beginning of the twentieth century, the men who looked hard and long at the logical bases of mathematics were to find the *Elements* riddled with fallacies and unstated assumptions. Yet the *Elements* remain, less perfect than they originally appeared to their compiler's contemporaries, frankly imperfect by the rigorous standards of modern mathematics, but still on the throne. For every domain of mathematics today is ruled by the axiomatic method, the system of Euclid's *Elements.*

Twenty-three hundred years after the Greek Eukleides lived and taught on the shores of the Mediterranean, mathematicians and scientists from all over the world gathered in Berkeley, California, under the shadow of the cyclotron, for a week-long international symposium on the axiomatic method and its relation to modern science. The ladder to truth was set on a far different shore, but the rungs were still the same: Definitions. Axioms. Theorems. Proofs.

FOR THE READER

Euclid's algorithm is one of the oldest techniques in arithmetic, probably even older than Euclid.

To find the greatest common divisor of two numbers

a and b by this method, we divide the smaller a into the larger b. If we obtain a remainder c, we divide c into a and so on, the remainder d being divided into c, e into d. Eventually we shall come to one of two possible situations:

1. Our division comes out even, in which case our last positive remainder is the greatest common divisor of the two numbers; or

2. Our remainder is 1, in which case the two numbers are relatively prime and their greatest common divisor is 1.

Both of these situations are illustrated in the simple examples below, to find the g.c.d. of 26 and 94 and of 26 and 101.

```
        3                                    3
  26 )94                            26 )101
     78    1                           78    1
     ‾‾                                ‾‾
     16 )26                            23 )26
        16    1                           23    7
        ‾‾                                ‾‾
        10 )16                            3 )23
           10    1                          21    1
           ‾‾                               ‾‾
            6 )10                           2 )3
              6    1                           2
              ‾                               ‾
              4 )6           g.c.d. = 1       1
                 4    2
                 ‾
                 2 )4
                    4
                    ‾
g.c.d. = 2          0
```

The reader may now enjoy using this same method to find the greatest common divisor for some larger pairs:
116 and 280; 507 and 1862; 280 and 882; 2475 and 19404

3

*The
Inexhaustible
Storehouse*

Mathematics began with these two basic concerns, and for centuries the subject was defined simply as "the science of form and number." Yet form has never been completely distinct from number. When, after the discovery of the irrational, mathematics found itself forced into the guise of form, it did not leave number behind. The beginnings of what we know today as the *theory of numbers* lie in Books VII, VIII and IX of Euclid's *Elements*.

The theory of numbers, or the *higher arithmetic* as it is often called, limits itself entirely to the whole numbers 0, 1, 2, 3, . . . and the relationships that exist among them. These numbers are a very simple sequence, formed by making each member one unit larger than the one that precedes it and continuing without end. They have challenged the minds of men for centuries because under their simple surface characteristics lie layer after layer of increasingly complex and utterly unexpected relationships.

This challenge was felt by Euclid. It has been felt, regardless of their individual specialties, by almost all the mathematicians who have followed him.

"The higher arithmetic," wrote

30

Karl Friedrich Gauss (1777-1855), known today and in his own lifetime as the Prince of Mathematicians, "presents us with an inexhaustible storehouse of interesting truths—of truths, too, which are not isolated, but stand in the closest relation to one another and between which, with each successive advance of the science, we continually discover new and wholly unexpected points of contact."

In this chapter we shall try to glimpse some of the treasures of this inexhaustible storehouse by examining a few of the mathematically interesting relationships which exist between two kinds of numbers—the primes and the squares. Both the primes and the squares were studied extensively in the *Elements* of Euclid; yet mathematicians are still discovering—in the words of Gauss—"new and wholly unexpected points of contact" between them.

Although the classification into *even* and *odd* is the most ancient, the most mathematically suggestive classification of the whole numbers greater than 1 is into those which can be divided by some number besides themselves and 1 (called *composite numbers*) and those which can be divided only by themselves and 1 (called *prime numbers*). The first few prime numbers are easily recognizable, for they are those the units of which cannot be arranged except in straight lines:

2	00	7	0000000
3	000	11	00000000000
5	00000	13	0000000000000

The units of all other, *composite* numbers can always be arranged into rectangles as well as straight lines:

31

4	00 00	9	000 000 000
6	000 000	10	00000 00000
8	0000 0000	12	000000 0000 000000 or 0000 0000

It is difficult to believe that no matter how high we go among the numbers, we shall continue to find numbers that can be arranged *only* in straight lines. Yet in Book IX (Prop. 20) of the *Elements,* Euclid proved that these essentially indivisible numbers—the primes—are infinite.

Euclid's proof is, of course, distinctly geometric in flavor. His numbers are straight lines, "beginning from a unit and in continued proportion," and his primes are lines "measured by the unit alone." The truth that he establishes, however, is the one above all others which makes numbers so interesting.

Euclid's proof rests upon the fact that if we multiply together any group of prime numbers, the number which is 1 more than the number we get as our answer will be either (1) another prime not in our original group or (2) a composite number which has, as one of its factors, a prime not in the group of primes we multiplied. This is because all of the primes we have multiplied must leave a remainder of 1 when divided into this next number:

$$2 \times 3 \times 5 = 30 \qquad \text{(30 + 1) divided by 2, 3 or 5 leaves a remainder of 1}$$

Euclid showed, therefore, that it would be impossible to

32

have a finite set which contained *all* the primes because by multiplying them and adding one to our answer we could always produce a prime not in our set of "all."

The relationship which exists between the divisible composite numbers and the indivisible primes is such a key to unlocking the secrets of numbers that the theorem which expresses it is universally acclaimed the Fundamental Theorem of Arithmetic.

Before stating this theorem, let us recall that by definition every composite number is divisible by some number other than itself and 1. This number which divides it must be prime or composite and, of course, smaller than the original. If it is composite, it must be divisible in turn by some number other than itself and 1, and so on. This process ends only when we come to a number which is not divisible by any other: a prime factor of the original composite number. It follows, then, that every composite number can be produced by the multiplication of primes or, conversely, can be factored into primes.

The Fundamental Theorem of Arithmetic states simply that this prime factorization for any composite number is *unique.*

This means that when we reduce a number like 36 to its prime factors $(2 \times 2 \times 3 \times 3)$, we know that although it has other factors $(4 \times 9$, for instance, and $6 \times 6)$, it can be reduced to *no other combination* of *prime* factors. By the Fundamental Theorem we know that the same thing will be true of a number like 18,674,392 —or any other number, no matter how large. We can thus work with any number n as a unique individual among the numbers. Not only do we know that it has a unique place in the sequence of numbers (between $n-1$ and $n+1$), but also we know that it is a unique combination

33

of certain prime factors $p_1{}^{k_1}p_2{}^{k_2} \ldots p_r{}^{k_r}$, where the p's represent different primes, and the k's how many times each prime appears as a factor.

The numbers which, next to the primes, have received the most attention from mathematicians are the squares. Their name comes to us from the eye-minded Greeks who noted that the units of a number when multiplied by itself always form a perfect square. They also noted something else of great interest about these squares when they were built up by successive borders of units:

```
0    0 0    0 0 0    0 0 0 0    0 0 0 0 0
     0 0    0 0 0    0 0 0 0    0 0 0 0 0
1           0 0 0    0 0 0 0    0 0 0 0 0
                     0 0 0 0    0 0 0 0 0
     1+3             0 0 0 0    0 0 0 0 0
     =4     1+3+5
            =9       1+3+5+7
                     =16        1+3+5+7+9
                                =25
```

Between the primes and the squares there are many interesting "points of contact," deep, intricate and completely unexpected. Yet the primes and the squares are basically very different numbers.

On page 35 we have printed a table of the first fifty numbers in each classification. Let us first examine only the last digits of these numbers. Among the squares we see immediately that not one of them ends in 2, 3, 7 or 8; in fact, the last digits follow a pattern—0, 1, 4, 9, 6, 5, 6, 9, 4, 1—which repeats indefinitely. Since, when we multiply a number by itself, the last digit of the product depends only upon the last digit of the number being multiplied, any number ending in 3 will have a square ending in 9, and so on. Obviously, there are infinitely many squares ending in each of the digits 0, 1, 4, 5, 6, 9 and none what-

0	100	400	900	1600
1	121	441	961	1681
4	144	484	1024	1764
9	169	529	1089	1849
16	196	576	1156	1936
25	225	625	1225	2025
36	256	676	1296	2116
49	289	729	1369	2209
64	324	784	1444	2304
81	361	841	1521	2401

THE FIRST FIFTY PRIME NUMBERS

2	31	73	127	179
3	37	79	131	181
5	41	83	137	191
7	43	89	139	193
11	47	97	149	197
13	53	101	151	199
17	59	103	157	211
19	61	107	163	223
23	67	109	167	227
29	71	113	173	229

soever ending in 2, 3, 7 or 8. But when we examine the last digits of the primes, we find that aside from 2 and 5 all primes end in 1, 3, 7 or 9. Since all even numbers are by definition divisible by 2 and all numbers ending in 5 divisible by 5, it is apparent that primes can end only in

1, 3, 7 or 9. But the primes, unlike the squares, are very unpredictable in their appearance among the numbers. We know by Euclid's proof that the number of primes is infinite, but are there—as with the squares—infinitely many primes ending in each of the possible digits?

The answer is given affirmatively by a very deep theorem proved over a hundred and fifty years ago by P. G. Lejeune Dirichlet (1805-1859). He showed that every arithmetic progression of numbers

$$a, a + d, a + 2d, a + 3d, a + 4d, a + 5d, \ldots$$

contains infinitely many primes when a and d have no common factor. If we take $a = 1$, 3, 7 or 9 (the only possible endings for primes) and $d = 10$, we know that in each of the four resulting progressions there are infinitely many primes: infinitely many primes ending in 1; infinitely many ending in 3; infinitely many ending in 7, and infinitely many ending in 9.

1, **11**, 21, **31**, **41**, 51, **61**, **71**, 81, 91, **101**,...
3, **13**, **23**, 33, **43**, **53**, 63, **73**, **83**, 93, **103**,...
7, **17**, 27, **37**, **47**, 57, **67**, 77, 87, **97**, **107**,...
9, **19**, **29**, 39, 49, **59**, 69, **79**, **89**, 99, **109**,...

If we look again at our table of primes and squares, we can see that it is no problem to write down the next entry in the column of squares: we simply multiply 50 by 50 and put down 2500.* But, to make the next entry in the column of primes, the best we can do is to examine the next odd number, 231. By inspection we see that it is divisible by 3, so we move on to the next odd number, 233. We try to divide it, in turn, by 3, 5, 7, 11 and 13 (all the primes which are less than its square root) and since none divides it we can conclude that it is prime, and write it

* We can also add together the first fifty odd numbers.

down as our next entry. This is the *only* general method for finding out whether a given number is prime.*

The classifications of the numbers which we have mentioned so far—even, odd; prime, composite; square and non-square—are so obvious that even if we do not usually think of all of them by name we cannot remember when we were not aware of them. Yet, among these groups of numbers there exist, in the words of the great Gauss, "wholly unexpected" points of contact. On the surface we have a not-unexpected relationship between the prime numbers and the odd. All the primes with one exception are odd, since every even number is by definition divisible by the only even prime, 2. When we separate the odd primes on the basis of their remainders when divided by 4, all are either of the form $4n + 1$ or $4n + 3$. Certainly we have no particular reason for expecting that these primes, falling into two mutually exclusive groups because of their relation to the first non-trivial square number, should present us with any significant and unvarying difference in their relation to the squares. Yet they do. This difference becomes apparent when we attempt to represent each of the first few primes as the sum of two squares. With 3, 7, 11, 19, 23, 31, and 43, we have no success at all; but we find that

$$5 = 1^2 + 2^2$$
$$13 = 2^2 + 3^2$$
$$17 = 1^2 + 4^2$$
$$29 = 2^2 + 5^2$$
$$37 = 1^2 + 6^2$$

and so on.

It is immediately suggested that every prime of the

* The largest known prime at the date of writing is $2^{9941} - 1$, found prime by D. B. Gillies, on Illiac II at the University of Illinois, April 21, 1963.

form $4n + 1$ can be represented as the sum of two squares, while not one prime of the form $4n + 3$ can be so represented. The theorem which expresses this relationship is even more specific, for it further states that the $4n + 1$ primes can be represented as the sum of two squares *in only one way*. This is the classic Two Square Theorem of Pierre Fermat. Although it involves no mathematical concepts which are not familiar to a bright child, it expresses a profound point of contact among the numbers, and one of the most "beautiful" relationships in all number theory.

Fermat wrote to a fellow mathematician that he had proved the Two Square Theorem by what he called "the method of infinite descent." He began with the assumption that there existed a prime of the form $4n + 1$ which could not be represented as the sum of two squares; proved that if there were such a prime, there would have to be a smaller prime of the same form which could not be so represented; and continued in this way until he got to 5, the smallest prime of the form $4n + 1$. Since 5 can be represented as the sum of two squares, the original assumption was false; the theorem, as stated, was true. The extreme difficulty of this proof can be grasped from the fact that although Fermat detailed it roughly to the extent we have here, it was not until almost a hundred years after his death that a mathematician was actually able to prove the Two Square Theorem.

In addition to the Two Square Theorem, we have a Three Square Theorem and a Four Square Theorem, both of which reveal interesting relationships between the square numbers and all the numbers. Both theorems deal with the same relationship, the representation of numbers as the sums of squares; but the Three Square Theorem penetrates much more deeply into the relationship than the Four Square Theorem.

FOUR SQUARE THEOREM: *Every number can be represented as the sum of four squares.*

There is no better example in number theory of the fact that it is easier to state a truth than to prove it. A little computation is enough to suggest that four squares are probably sufficient to represent any number. The fact was probably known in the early years of the Christian era. It was then restated as part of a more general theorem, and proved by Fermat. Although Fermat remarked in a letter to a friend that no proof had ever given him more pleasure, he neglected to reveal the details to anyone, and the proof died with him. Leonhard Euler (1707-1783), one of the greatest, and certainly the most prolific mathematician who ever lived, then tackled the part of Fermat's theorem pertaining to the squares. In fact, off and on, he devoted forty years of his long life to it—without success. Eventually, though, with the help of much of the work which Euler had done, the Four Square Theorem was proved by Joseph Louis Lagrange (1736-1813). A few years later Euler brought forth a more simple and elegant proof than Lagrange's of the theorem which had caused him so much difficulty, and it is now the proof generally followed.

For such representation of all numbers as the sum of four squares, we rely extensively upon the use of the square of 0, particularly in the case of those numbers which are squares to begin with or those numbers, like the primes of the form $4n + 1$, which are the sum of two squares. It is obvious from these that four squares are not by any means necessary to represent every number as the sum of squares. The question which then occurs is whether or not we can determine, by any general rule, the particular group of numbers for which four squares are necessary. This is exactly the answer which the Three Square Theorem gave. There is, according to the theorem, a particular group of numbers, the first of which is 7, that cannot be represented by any fewer than four squares; for all other numbers, three squares are sufficient.

THREE SQUARE THEOREM: *Every number can be represented as the sum of three squares except those numbers of the form $4^a(8b + 7)$.* *

Now the Four Square Theorem is by no means trivial. Although the representation of the smaller numbers as the sum of four squares is easy to perceive, there is no assurance that as the numbers get larger more squares will not be required. Yet, when compared to the Three Square Theorem, which pinpoints the specific type of number (and not an obvious, straightforward type, either) requiring four squares for representation, the Four Square Theorem is distinctly inferior—"much less deep," in the opinion of mathematicians.

To discover such deep relationships among the numbers, we must not look at them with jaded eyes. Youth, freshness, and perhaps mentally standing on one's head help. We also need a gift for seeing such relationships.

There is a relationship between the squares and the odd primes which is even more mathematically exciting than the one Fermat expressed in the Two Square Theorem, fully as deep as if not deeper than the relationship expressed in the Three Square Theorem. But it would not even be observed by anyone who did not have the gift. Although this particular relationship had been observed earlier, the young Gauss (he was eighteen at the time) discovered it wholly on his own and was delighted with it. To him it was always the Gem of Arithmetic. More formally, it is known as the Law of Quadratic Reciprocity (quadratic meaning simply "of or pertaining to the squares").

* It was proved by Gauss.

The Law of Quadratic Reciprocity deals exclusively with the same kinds of numbers as does the Two Square Theorem of Fermat—the squares and the primes classified according to the remainders they leave when divided by 4. Let p and q be any pair of odd primes; there exists a beautiful and delicately balanced relationship between these two apparently unrelated problems:

1. To find an x such that $x^2 - p$ is divisible by q.
2. To find a y such that $y^2 - q$ is divisible by p.

According to the Law of Quadratic Reciprocity, both problems are solvable or both unsolvable unless both p and q leave a remainder of 3 when divided by 4, in which case one of the problems is solvable and the other is unsolvable.

"The mere discovery of such a law was a notable achievement," writes E. T. Bell in *Men of Mathematics*. "That it was first proved by a boy of nineteen will suggest to anyone who tries to prove it that Gauss was more than merely competent in mathematics."

It took Gauss a year to prove the Law of Quadratic Reciprocity. "It tormented me and absorbed my greatest efforts," he wrote later. His was the first proof of this beautiful law and he published it proudly in the *Disquisitiones Arithmeticae* under the title of Fundamental Theorem. But he was not at all satisfied with his proof: ". . . it proceeds with laborious arguments and is overloaded with extended operations." In the next seven years he proved the Law of Quadratic Reciprocity in four more ways, using completely different principles. The first three of these four proofs, all of which he conceded were logically satisfactory, he dismissed as "derived from sources much too remote." The last he published with the frank statement, "I do not hesitate to say that till now a *natural* proof has not been produced. I leave it to the authorities to judge whether [this] proof which I have recently been fortunate enough to discover deserves this description."

The "authorities" apparently decided that it did, for this fifth proof (known as "the third" because it was the third one he published) is the proof which is universally used today. But Gauss himself could not have been satisfied: three *more* times in his life he proved the Law of Quadratic Reciprocity, his Gem of Arithmetic.

Lest we feel at this point that Gauss himself may have singlehandedly exhausted the inexhaustible storehouse of interesting truths which he found the natural numbers to be, we might mention that he went on to tackle the problem of *biquadratic* reciprocity where x and y are taken to the fourth power. A by-product of his solution was the creation of the theory of algebraic numbers, which we shall touch on in Chapter 7. Perhaps it is too much to mention that the general case of x and y taken to the nth power still remains in the storehouse!

It is curious that we usually think of arithmetic as the exact science, the science of right answers, the cut-and-dried science. But that is because we are thinking of the arithmetic of the elementary school, not the "Queen of Mathematics." In elementary arithmetic we perform operations on the numbers, first with accuracy, and then with speed. The ideal is most nearly achieved by the great electronic computers which, in spite of the awe they generate, can do no more difficult arithmetic than a high school boy or girl who is well trained; they can, however, do it faster and more accurately. An electronic computer is a mere drudge of the Queen of Mathematics. Although even Gauss loved to compute, he never failed to perceive the queen's real challenge.

"The questions of the higher arithmetic," he wrote, "often present a remarkable characteristic which seldom appears in more general analysis and increases the beauty of the former subject. While analytic investigations lead to the discovery of new truths only after the fundamental principles of the subject (which to a certain degree open the way to these truths) have been completely mastered, on the contrary in arithmetic the most elegant theorems frequently arise experimentally as the result of a more or less unexpected stroke of good fortune,* while their proofs lie so deeply imbedded in the darkness that they elude all attempts and defeat the sharpest inquiries. Further, the connection between arithmetical truths, which at first glance seem of widely different nature, is so close that one not infrequently has the good fortune to find a proof (in an entirely unexpected way and by means of quite another inquiry) of a truth which one greatly desired and sought, in vain, in spite of much effort. These truths are frequently of such a nature that they may be arrived at by many distinct paths and that the first paths to be discovered are not always the shortest. It is therefore a great pleasure, after one has fruitlessly pondered over a truth and has later been able to prove it in a roundabout way, to find at last the simplest and most natural way to its proof." †

Today, twenty-five hundred years after the Pythagoreans first perceived that the squares and the primes are very interesting numbers, there are still many questions to be answered about their relationship to one another. Is there, for instance, a prime between every pair of consecutive squares? Are there infinitely many primes that are just one unit greater than a square $(x^2 + 1)$?

The inexhaustible storehouse awaits.

* It is interesting to note that Gauss first observed the Law of Quadratic Reciprocity when he was computing the decimal representation of all reciprocals through $\frac{1}{1000}$ in an attempt to find a general rule for determining the period of a repeating decimal.

† The quotations from Gauss are translated from the Latin by D. H. Lehmer and appear in David Eugene Smith's *A Source Book in Mathematics.*

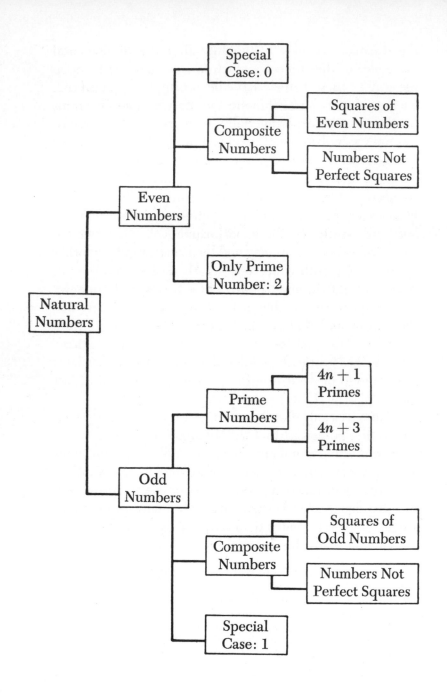

FOR THE READER

The squares are numbers it is fun to play with by eye, as the Greeks played with them. If the reader will provide himself with a set of counters or just a sheet of graph paper, he will find it fun to try to determine why the following system of multiplication—which is achieved by addition, subtraction and division of squares—*works:*

To multiply 7 by 6—

we take the sum of 7 and 6,	$7 + 6 = 13$
square it,	$13^2 = 169$
subtract the square of 7,	$169 - 49 = 120$
subtract the square of 6,	$120 - 36 = 84$
divide by 2,	$84 \div 2 = 42$
	$42 = 7 \times 6$

Why does it work?

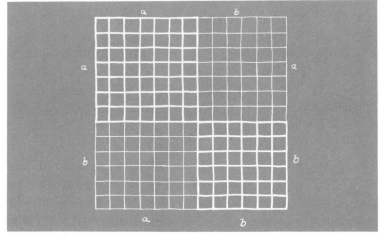

4

*A Number
for Every Point
on the Line*

FROM WHOLE NUMBERS TO RATIONAL numbers to irrational numbers . . .

This step-by-step extension of the idea of Number was forced upon mathematics by the contemplation of so seemingly simple a thing as the straight line. Faced with the fact that the straight line which is the diagonal of the unit square can be measured with truly mathematical accuracy only by the square root of 2, the Greeks concluded that *there was no such number.* The men who followed them, however, have insisted upon the fact that *for every point on the line, there must be a number.*

It would seem that even in two thousand years such diametrically opposed points of view could never be brought together. Yet if we begin at both ends of the time—in the third century B.C. and the nineteenth century A.D.—we find that the Greek solution of this problem, which was a revised theory of proportion, and the modern solution, which is the concept of the real number line or *arithmetic continuum,* are in essence the same.

The revised theory of proportion which allowed Greek mathematics to move forward again, although in the guise of Form now rather than of Number, is contained in the fifth and sixth books of the *Elements* and is considered without question the finest mathematics in Euclid. It is, almost entirely, the work of Eudoxus.

Eudoxus was a poor young student who walked every day to Athens to sit at the feet of Plato. His genius was recognized and he became eventually a great and honored teacher himself, with many personal achievements in astronomy and geometry. His masterpiece was his theory of proportion and, specifically, his redefinition of "in the same ratio" so that it could be applied to the newly discovered incommensurable magnitudes as well as to the traditional commensurable magnitudes.

Under the universal rule of Number, before the discovery of the irrationality of $\sqrt{2}$, ratio had been conceived by the Pythagoreans as the expression of the relative magnitude of two whole numbers, or lengths. We might think that the need for such rational expressions arose practically in measurements where the distance to be measured fell between two units, or whole numbers. Actually, it arose as a result of Pythagorean interest in the purely theoretical relationship between magnitudes.

Given two magnitudes like A and B below, how can we express the relationship between them in whole numbers?

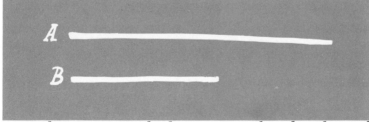

A simple way is to multiply, or repeat, these lengths until we reach a point where both totals coincide. In the example above, if we take five of the length labeled A and nine of the length labeled B, we will find that we have two equal lengths. Since $9B = 5A$, the relative magnitude of A to B is 9 to 5, or the "rational number" 9/5. (We must use quotation marks here, for Greek mathematicians from Pythagoras to Diophantus (A.D. 300) did not consider these rational expressions to be numbers. As far as they were concerned, the only real numbers were still the whole numbers.)

Another way of determining relative magnitude (also known to the Pythagoreans) is the method we still use today to find the greatest common divisor of two numbers—Euclid's algorithm. If we measure off A by B and then measure off B by the remainder C, we obtain eventually a remainder (in our example: D) which exactly measures the previous remainder.

It is easy to see that D measures both A and B exactly, A 9 times and B 5 times. Taking D as the common unit, the relative magnitude of A and B is, as we also found by our first method, 9 to 5—the "rational number" 9/5;

$$\frac{A}{B} = \frac{9}{5}$$

or, in the familiar language of proportion, A is to B as 9 is to 5 ($A:B :: 9:5$).

This definition of ratio is perfectly adequate if we wish to express the relative magnitude of the base and the hypotenuse of the ancient 3-4-5 right triangle pictured on the left below. But if we try to use the same methods to find the ratio between the unit base of an isosceles right triangle on the right and its hypotenuse, we are in trouble. No matter how many times we take the hypotenuse and how many times the side, we will never reach a point where our totals coincide. If we try to use the method of finding the greatest common measure, which worked so well for *A* and *B* above, we will never obtain a remainder which is exactly contained in the preceding remainder.*

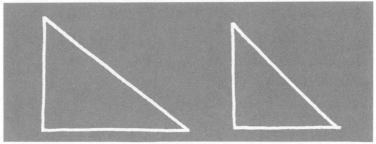

In geometry we say that these two lengths are incommensurable. In arithmetic, if the only numbers we have are the whole numbers, we cannot express the relationship between them.

Yet our eye tells us that the base and the hypotenuse of the triangle on the right have—like any two lines—a relative magnitude, even if we cannot express it in the only numbers we have. It is the "ratio" of 1 to $\sqrt{2}$; but we cannot *call* this a ratio as long as ratio is defined in the traditional sense of relative magnitude expressed by whole numbers.

Eudoxus solved this difficulty like a true mathematician. He simply redefined ratio so that it could be applied to incommensurables as well as to commensurables.

* If we appear to be successful, it is only because of an error introduced by the thickness of our lines.

It was this new definition which Euclid used in the *Elements*.

Eudoxus limited his definition of ratio to finite magnitudes of the same kind. He then proceeded to the crux of the matter. What do we mean when we say that magnitudes are "in the same ratio"?

The simplest way to determine that a/b and c/d are in the same ratio is to reduce them to lowest terms. We say that $\frac{4}{6}$ and $\frac{8}{12}$ are the same since both when reduced to lowest terms are the fraction $\frac{2}{3}$. A more complicated way, but the one more appropriate to Eudoxus' definition of "in the same ratio," is the following:

We say that a/b and c/d are in the same ratio when we can multiply the numerators a and c by some whole number m and the denominators b and d by some whole number n so that

$$ma = nb$$

and

$$mc = nd$$

To make this process clearer, let us determine by this method whether $\frac{4}{6}$ and $\frac{6}{9}$ are in the same ratio. We multiply both numerators by the same number (3) and both denominators by the same number (2):

$$3 \times 4 = 12 \qquad 3 \times 6 = 18$$
$$2 \times 6 = 12 \qquad 2 \times 9 = 18$$

If by such multiplication with whole numbers there is some way we can make our new numerators equal, respectively, to our new denominators, we say our two original ratios are in the same ratio. This, we say, is what we mean by "in the same ratio."

50

The difficulty is that our method is applicable only to ratios of whole numbers; in other words, commensurables. When we are dealing with the ratios of commensurables, we can always find an m/n such that

$$ma = nb$$

and

$$mc = nd$$

but when we are dealing with incommensurables, ma and mc will always be greater than nb and nd, or less. They will never, no matter what m/n we select, be exactly equal. However—and this is the fact that Eudoxus seized upon for his masterly redefinition—if

$$\frac{a}{b} = \frac{c}{d}$$

and if we multiply the numerators by the same whole number m and the denominators by the same whole number n, regardless of whether we are dealing with commensurables or incommensurables, we can never get the result that ma is greater than nb while mc is less than nd. If our original ratios are actually in the same ratio, the numerators of our new ratios will always *both* be greater or *both* be less than our new denominators, or the new numerators and the new denominators will be, respectively, equal.

This then, said Eudoxus, is in essence what we mean by "in the same ratio."

In modern notation we can state this definition as follows:

Consider a/b and c/d. If we multiply a and c by the same number m, and b and d by the same number n, and if we obtain one of the following situations, and no other:

$$ma > nb \qquad \text{and} \qquad mc > nd$$

or

$$ma = nb \qquad \text{and} \qquad mc = nd$$

or

$$ma < nb \qquad \text{and} \qquad mc < nd$$

then $a/b = c/d$.

Unfortunately, when in the fifth definition of Book V of the *Elements* Euclid had to present this definition of "in the same ratio," he, like Eudoxus, did not have the benefit of algebraic notation; and he had to write: "Magnitudes are said to be in the same ratio, the first to the second and the third to the fourth, when, if any equimultiples whatever be taken of the first and third, and any equimultiples whatever of the second and fourth, the former equimultiples alike exceed, are alike equal to, or alike fall short of the latter equimultiples, respectively, taken in corresponding order."

It was this definition which Isaac Barrow (1630-1677), who voluntarily gave up his professorship at Cambridge to the young Newton, called "that Scare-crow at which the over modest or slothful Dispositions of Men are generally affrighted." *

Today, more than two thousand years since Eudoxus formulated this definition, it is echoed almost word for word in the modern definition of *equal numbers,* which,

* He went on to add: "They are modest, who distrust their own Ability, as soon as a Difficulty appears, but they are slothful that will not give some Attention for the learning of Sciences; as if while we are involved in Obscurity we could clear ourselves without Labour. Both of which Sorts of Persons are to be admonished, that the former be not discouraged, nor the latter refuse a little Care and Dilligence when a Thing requires some Study." (We recommend to the reader the words of Isaac Barrow.)

in very much the same way that Eudoxus' definition of "in the same ratio" enabled the Greek mathematicians to deal with incommensurable lengths, enables modern mathematicians to deal with irrational *numbers*. There is, however, high irony in this. When the Greeks found that there were points on the line for which their mathematics had no exact numerical expression, they fled from Number into Form and took sanctuary in a geometric theory of proportion which could handle incommensurables. Yet in this same sanctuary, although they never found it, was the saving concept of number which they sought—*a unique number for every point on the line.*

In the two thousand years that elapsed between the Greek theory of proportion and the modern concept of the arithmetic continuum, the irrational numbers led a curious "here and not here" existence. They were mostly "not here" until the late sixteenth century. At that time the decimal notation began to come into common use, and mathematicians to their delight saw rational and irrational numbers fall into place like well-ordered regiments!

All decimals can be thought of as never-ending representations of numbers.

Some, after a certain point, repeat 0's indefinitely:
like ½, or .50000000000000000000000 . . .
Some repeat another single digit:
like ⅓, or .33333333333333333333333 . . .
Some repeat after a certain point a series of digits:
like ⅐, or .$\overline{142857}$1428571428571428 . . .
Some never end and never repeat:
like π, or 3.1415926535897932384626 . . .

It is very easy to show that all rational numbers in their decimal representation will repeat; and, conversely, that all repeating decimals are representations of rational numbers.

Consider the rational number $\frac{1}{17}$. To obtain a decimal representation, we simply divide 17 into 1. Sometime —within the first 16 steps of this division—we must obtain a remainder which we have obtained before, since there are only 16 possible positive remainders. When we do, our quotient must of necessity begin to repeat. In the case of $\frac{1}{17}$, the decimal representation actually does have a 16-place period:

$$.\overline{0588235294117647}058823 \ldots$$

We can say, then, in general terms that the decimal representation of any rational number a/b will repeat within $(b-1)$ decimal places.

Now let us consider the reverse situation where we are given a repeating decimal and wish to obtain a rational representation of it. We take, for example, the repeating decimal .1212121212121212121. . . . We multiply this decimal by 100 so that we have a whole number 12 followed by the repeating decimal .121212121212121212121. . . . We then subtract our original repeating decimal, which is the same as this same decimal tail:

$$
\begin{array}{r}
12.121212121212121212121 \ldots \\
- \quad .121212121212121212121 \ldots \\
\hline
12.000000000000000000000 \ldots
\end{array}
$$

What we have done is to subtract our original decimal from a number 100 times as great. The answer of 12 which we obtain is thus equal to 99 (i.e., $100-1$) times our original decimal. The original decimal, therefore, must be equal to 12 divided by 99, or $\frac{4}{33}$, which is its representation as a rational number reduced to lowest terms. Again, it is clear that we can always do exactly this: we can always obtain for any repeating decimal an expression as a rational number.

Since, as we have seen above, all rational numbers can be represented as periodic decimals and since all periodic decimals represent rational numbers, it follows that all irrational numbers can be represented by non-repeating decimals and that all non-repeating decimals represent irrational numbers. Granted that this is not a very precise definition of *irrational number,* it was so much more numerical than anything mathematicians had seen since the Pythagorean discovery of the irrationality of $\sqrt{2}$ that they welcomed it without question. They proceeded to apply the operations of arithmetic to these new numbers in the same manner they applied them to the whole numbers and the fractions, and nobody worried much about the niceties.

In the late nineteenth century, all of this was changed. Certain mathematicians, including Richard Dedekind (1831-1916) and Georg Cantor (1845-1918), saw the necessity for a truly precise formulation of what mathematicians call the *real* numbers, the numbers for the points on a line. (The reason for this name will become clear in Chapter 7.) Curiously enough, at that time, twenty-three hundred years after Euclid compiled the *Elements,* they expressed their ideas not in the terms of some recent mathematical development but very much in the terms of the Eudoxian theory of proportion as presented in the fifth book of the *Elements.*

Although there are today several ways in which irrational numbers can be precisely defined, the most popular definition remains that of Dedekind and bears the dramatic title of "a Dedekind cut." Dedekind formulated this definition with full modern rigor, but we can grasp it more easily if we discuss it in a rather rough fashion, relying heavily on our intuitive understanding of "number" and "line." We begin by thinking of all the rational numbers as being paired off on a line with those points which they represent—all lengths being measured from an arbitrary origin point labeled 0. For simplicity's sake, we can concern ourselves now with only that part of the line which is to the right (or positive side) of 0:

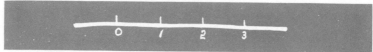

From an everyday point of view, although this line looks as familiar as an ordinary ruler, there are several rather unusual things about it. For example, it has no beginning and no ending. If we select on it any two points which have been paired with rational numbers, we can always find between these as many more points—or numbers—as we please. Say that we select two points as "close" as $\frac{2}{1000}$ and $\frac{3}{1000}$. Between these two points lies the point $\frac{5}{2000}$. Between $\frac{5}{2000}$ and $\frac{3}{1000}$ lies $\frac{8}{3000}$, and so on. In general, if we take any two rational numbers a/b and c/d, add their two numerators and add their two denominators, we shall obtain a rational number:

$$\frac{a + c}{b + d}$$

which lies between them. There is no "nextness" among these rational numbers.

Now let us make our first cut in this line. Let us say that we will cut it at the point which is paired with the rational number ½. The complete line will have been cut into two pieces which together include every point, or rational number, on the entire line. If our cut line is to be in just two pieces, the cut point ½ must be in one piece or or the other; it cannot, of course, be in both:

56

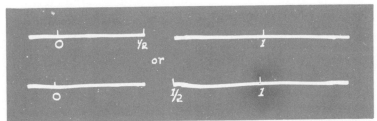

If ½ is included, as in the upper example, in the left-hand segment of the line, it must be the largest number on that side. Since there is no "nextness" among the rational numbers, the right-hand segment of the line can have no smallest number. But if ½ is included, as in the lower example, on the right-hand side of the cut, then it must be the smallest number on that side and the left-hand side now can have no largest number.

Thinking in this manner, we are defining rational numbers as *cuts* which divide the line into two parts in such a way that one—and only one—of the parts has either a largest or a smallest number. This is a curious enough definition; but like Alice, we find that things become curiouser and curiouser.

What if we break the line at a place where there is no point and, hence, no rational number? This is completely possible, because although the rational numbers are dense upon the line, they are not continuous. If they were, and if there were a rational number for every point on the line, the Pythagoreans would have found a rational number which exactly measured the square root of 2. What happens when we cut the line at the place where there "should" be a number-point equal to $\sqrt{2}$? The line divides, as before, into two parts but with an important difference:

There is now no largest rational number in the left-hand part and no smallest rational number in the right-hand part.

Such a cut—according to Dedekind's definition—*is* an irrational number!

In a much more formal and precise statement, the definition can be put in the following way:

An irrational number *a* is defined whenever the rational numbers are divided into two classes *A* and *B* such that every rational number belongs to one, and only one, class and (1) every number in *A* precedes every number in *B*, and (2) there is no last number in *A* and no first number in *B*; the definition of *a* being that it is the only number which lies between all numbers in *A* and all numbers in *B*.

We shall not here expand in detail upon the similarities between this definition and the Eudoxian definition of "in the same ratio." The reader who is particularly interested will find a complete statement in Sir Thomas Heath's edition of the *Elements*. Suffice it to say that the two definitions, separated by more than two thousand years of mathematical thought, are in essence the same.

From Dedekind's definition of an irrational number as a cut in the rationals, we can now proceed to the statement of the axiom upon which all of arithmetic, and hence all of mathematics, rests. If we replace every Dedekind cut in the rational numbers with a point and a number (a non-rational, or irrational, number) so that regardless of where on the line we make a cut, we shall always cut at a number-point pair, we can state what is known as the Cantor-Dedekind axiom:

It is possible to assign to any point on a line a unique real number, and, conversely, any real number can be represented in a unique manner by a point on the line.

We have come in some twenty-five hundred years

from the despairing conclusion of the Pythagoreans that for some lengths there were no numbers to the completely satisfying conclusion of Dedekind and Cantor that for every point on the line there is a number!

From whole numbers—to rational numbers—to irrational numbers! These, taken together, are the *real* numbers; and once again Number (now *real number*) rules the Universe.

FOR THE READER

It is fun to test for oneself the fact that every rational number can be represented as a repeating decimal, particularly when such rational numbers as those listed below are taken for the experiment:

$$\frac{1}{19}, \frac{1}{23}, \frac{1}{41}$$

It is also fun to test for oneself the fact that, conversely, every repeating decimal represents a rational number. The reader is urged to apply the method given on page 54 to such decimals as:

.88888888888888888 . . .
.10101010101010101 . . .
.23523523523523523 . . .
.18888888888888888 . . .
.28545454545454545 . . .
.14285714285714285 . . .

Then just for fun he should "make up" some repeating decimals for himself and discover what rational numbers *they* represent!

5

Journey That Begins at O

THE STRAIGHT LINE AND THE CIRCLE.

These two lines exercised such a fascination over the ancient mind that they limited the instruments of mathematical construction, determined the subject matter of most of the mathematics, and provided mathematical "problems" that were not to be finally disposed of for more than two thousand years.

Fascination with the straight and the round apparently blinded Greek eyes to the lines which they actually saw around them. Their geometry was based on an axiom which stated in essence that parallel lines never meet, their intuitive and undefined idea of a *straight* line being inextricably bound up with this axiom; yet the architects of the Parthenon built their pillars so that they bulged in the middle, for they knew that if they made the sides straight and parallel they would appear to curve in toward each other. They must have observed that the parallel sides of a straight roadway appear to converge as they approach the horizon. They must also have observed that circles always appear elongated except when the eye is on the axis of the curve. Nevertheless, the Greek mathematicians knew they *knew* that the parallel lines which appeared to meet could never by their very nature

60

meet and that circles, regardless of appearances, were in their actuality round. They limited themselves in their mathematics to the perfect essence of these figures, and did not concern themselves with the imperfect figures which their eyes saw all around them.

The only geometric constructions which the Greek mathematicians considered permissible, or "pure," were those made with a straightedge, an unmarked rule which was the mechanical equivalent of the straight line, and a compass, which was the mechanical equivalent of the circle. They then conceived that the "solution" of any geometric problem must be effected by these two instruments, alone.

This made things a lot harder. Problems which would yield gracefully to other instruments remained "unsolved" for two thousand years!

Undoubtedly the most dramatic of these ancient problems, which were to be with mathematics for so many centuries, was the problem of duplicating the cube. According to tradition, the people of Athens, suffering the ravages of a great plague, consulted the oracle at Delos. How could they placate the angry gods who had sent this plague upon them? The oracle replied that they should double the size of the cubical altar to Apollo. The obedient Athenians promptly built a cube with each side twice as long as that of the original and thereby produced an altar which was by volume eight times the size of the original altar. The gods, naturally enough, did not appear to be placated and the plague continued.

A century later, although the plague had long since run its normal course, the Greek mathematicians were still struggling with the problem of doubling the volume of a given cube. Since the unit cube has a volume of 1 (or $1 \times 1 \times 1$), a cube with volume twice as great must

be represented in modern notation by the formula
$$x^3 = 2$$
Solving for x then is the equivalent of extracting the cube root of 2.

The Greek mathematicians assumed that the problem, since it was proposed by the gods, required an exact answer and one which could be effected in its construction by the gods' chosen instruments, straightedge and compass alone. A rather familial picture of their difficulty is presented by Eratosthenes, who lived a while after Euclid and is famous for a surprisingly accurate measurement of the earth and for a "sieve" which is still the basic principle of all tables of prime numbers.

"While for a long time everyone was at a loss," Eratosthenes wrote, "Hippocrates of Chios was first to observe that if between two straight lines of which the greater is double the less it were discovered how to find two mean proportionals in continued proportion, the cube would be doubled; and thus he turned the difficulty of the original problem into another difficulty, no less than the former." *

Eratosthenes went on to report that Menaechmus, who was a pupil of Eudoxus, found two solutions to this problem, both effected by the intersection of conic sections. This is the first mention in mathematical literature of those beautiful and ubiquitous curves—the hyperbola,

* In modern notation we would state this problem of Hippocrates as:

To find $\frac{x}{y}$ such that $\frac{A}{x} = \frac{x}{y} = \frac{y}{2A}$, with A and $2A$ being two given straight lines.

From the equation $\frac{A}{x} = \frac{x}{y}$, we have $x^2 = Ay$. Squaring both sides of this equation, we obtain $y^2 = \frac{x^4}{A^2}$.

the parabola and the ellipse. To Menaechmus is given the credit for their discovery, although their names were given them much later by another mathematician.

Oddly enough, the names themselves go back as far as Pythagoras who, like the Greeks that followed him, never paid any attention at all to the generally elliptical appearance of the circle, the parabolic paths of projectiles, or the hyperbolic arches cast by shaded lanterns. One of the problems that did, however, interest Pythagoras was that of drawing upon a given segment a figure— triangle, square or pentagon—that was required to be the size of some other given figure of a different shape. In the course of the solution of this problem, one of three things might happen. The given line segment would be too short (*ellipsis*), exactly the right length (*parabole*) or too long (*hyperbole*). These same words have come very generally into English in the *ellipsis*, three dots which mark the omission of words; the *parable*, which tells one story but parallels another that although untold is the real story; and the extravagant exaggeration of statement which we call *hyperbole*. These names, suggesting as they do the arithmetical relations of *less than, equal to* and *more than*, were given to the conics by Apollonius of Perga, who followed Archimedes.

The conic sections of Menaechmus seemed to possess the Greek mind even though the interdiction against con-

---

From the equation $\frac{x}{y} = \frac{y}{2A}$, we obtain $y^2 = 2Ax$.

Substituting this value for y^2 in the preceding equation, we find that $\frac{x^4}{A^2} = 2Ax$, or $\frac{x^3}{A^3} = \left(\frac{x}{A}\right)^3 = 2$.

It follows, therefore, that the desired $\sqrt[3]{2}$ is $\frac{x}{A}$.

(The reader who finds his algebra rusty would do well to recall that Hippocrates had none at all to help him formulate his problem!)

struction by instrument other than straightedge and compass cast them out of the pure geometry of the day. The simplest way to produce the various sections is by cuts and cross-cuts of a solid circular double cone. (The reader may enjoy producing the various sections by cutting a cone of light with a piece of cardboard placed at varying angles.) A cut exactly parallel to the angle of the cone will give us a parabola. If the angle of our cut is within the angle of the cone, we obtain the two branches of the hyperbola; while if it is outside, we obtain an ellipse. A straight cut parallel to the base will give us the circle, or limiting form of the ellipse.

We can also think of each conic curve—ellipse, parabola, hyperbola—as the path of a point which must move according to certain rules which determine the curve it makes. This is most intuitively clear when we think of a circle as the path of a point which must always be a given distance (the radius) from another point (the center of the circle):

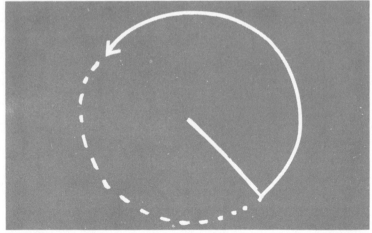

An ellipse is the path of a point which must always move so that the sum of the distances from two given points (called the foci) is always the same:

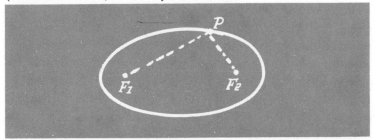

A hyperbola is the path of a point which must move so that the *difference* of the distances from two given points is always the same:

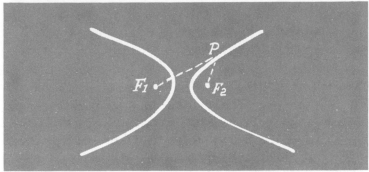

A parabola is the path of a point which is always the same distance from a given point that it is from a given line:

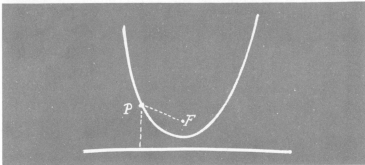

Euclid himself wrote a treatise on the conics, since lost. It was not Euclid, however, but Apollonius who developed the fundamental properties of the conics in remarkable generality. Apollonius first showed that all conics are sections of any circular cone, right-angled or oblique, and—in spite of the truly cumbersome expression available to him—gave for the first time the fundamental property of all conics, which we shall present later in modern notation.

This is the kind of achievement which any mathematician respects. It is the sort of work the English mathematician Littlewood was thinking of when he remarked to his friend Hardy that the Greek mathematicians were not just clever boys, scholarship candidates, but—in the language of his Cambridge—"fellows at another college."

Pappus, who has been called the worthiest commentator on the work of Apollonius, then showed that the ratio of distance of any point on any conic from a fixed point (the focus) and a fixed line (the directrix) is constant. This ratio, which we express as e, is called the "eccentricity" of the curve. A conic is an ellipse, a parabola or a hyperbola according as e is less than 1, equal to 1, or more than 1. In the circle e is 0.

Of such work by the ancient Greeks, a president of the British Association for the Advancement of Science wrote: "If we may use the great names of Kepler and Newton to signify stages in the progress of human discovery, it is not too much to say that without the treatises of the Greek geometers on the conic sections there could have been no Kepler, without Kepler no Newton, and without Newton no science in the modern sense of the term. . . ."

The earth follows a nearly elliptical orbit around the sun; projectiles approximate parabolic paths, a shaded light illuminates a hyperbolic arch. If it required the Delian problem to reveal these common curves to eyes blinded by circles, the problem would have done more than its turn for mathematics. But, indirectly, there were more great gifts to come.

It was some seventeen hundred years after Menaechmus, Apollonius and Pappus that an arrogant young Frenchman published a short mathematical treatise as a supplement to a larger philosophical work which he expected to ensure his immortality. This mathematical treatise, entitled *La Géométrie,* began with one of the most important sentences in the history of mathematics:

"Any problem in geometry," wrote René Descartes, "can easily be reduced to such terms that a knowledge of the lengths of certain lines is sufficient for its construction."

We have already seen how the discovery that the diagonal of the square is incommensurable with the side had driven the Greeks into a geometry without number and a theory of numbers expressed in the terms of geometry. "This great step backwards," it is regretfully called by Morris Kline in his *Mathematics and Western Culture.* Descartes, by applying the concept of the variable, which he took from algebra (the great Eastern contribution to mathematics),* to the ancient method of mapping by coordinates, which was known to the Babylonians and the Egyptians, reversed this step into a giant stride forward—a stride in fact into modern mathematics.

* The historical development of algebra has been characterized in three stages: (1) *rhetorical* algebra, in which problems were solved by a process of logical reasoning but were not expressed in abbreviations or symbols; (2) *syncopated* algebra, in which abbreviations and symbols were used for certain quantities and operations occurring most frequently; and (3) *symbolic* algebra, in which completely arbitrary symbols are used for all forms and operations—a development of the period immediately before and after Descartes.

It does not matter that in doing so, he was as much under the spell of utterly impractical geometric construction problems as were the ancient Greeks. The application of the method of coordinate mapping to geometry and algebra, which is called in the history of mathematics the *invention of analytic geometry* and credited to the young Frenchman, was one of those innovations which, as soon as they are finally made, seem as if they had always been inevitable. It freed both subjects from bonds which until then had appeared inherent in them. Geometrical figures were transformed into algebraic equations and equations into figures. Problems which had eluded the genius of the Greeks dropped into the hands of schoolboys.

In this chapter we shall offer a glimpse of this new tool by examining it in relation to the line and the circle of Euclid and the conic sections of Apollonius—in short, the curves known and studied geometrically by the ancient Greeks.

We begin with the selection on the plane of a point—it may be any point we care to choose—and we label this point "*O*" for *origin.* It is at this point that we set out on a very different mathematical road from that traveled by the Greeks.

Through the point *O* we draw a line which extends indefinitely to the right and to the left of *O*. On it, we mark off units very much as years are marked off from an origin point which is the birth of Christ. The units after the birth of Christ, or to the right of *O*, are labeled with a plus; the units before, or to the left of *O*, with a minus. We then draw perpendicular to our first line another which passes through *O* and extends indefinitely above and below it. On this line we mark off units as degrees of temperature, for instance, are marked off on the thermometer: the units "above" *O* being labeled with a plus and the units "below" with a minus. We call our original horizontal line the *x*-axis and this new vertical line which is perpendicular to it, the y-axis.

On the plane which we have marked off we can now locate *uniquely* any point by stating its position on the plane in the terms of its coordinates on the x and y axes. One number (x) tells its distance from the y-axis; another (y), its distance from the x-axis. This x, y pair is the address, as it were, of the point on the plane:

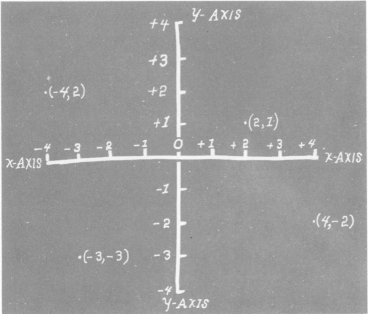

It is, of course, essential that the "addresses" of these points be stated in the proper order. Just as 70 Twenty-sixth Street is not the same address as 26 Seventieth Street, the number pair 26, 70 does not locate on the plane the same point that the number pair 70, 26 does. (If the reader will transpose the numbers in each pair located on the diagram above, he will find that he has located three entirely different points, one [—3, —3] remaining the same point even when the coordinates are transposed.)

The concept of points as such ordered number pairs is the key to analytic geometry. For this reason the coordinates of a point are always given in order: the x coordinate first, then the y coordinate. Our new definition of a point is a long way from Euclid's "A point is that which has no part":

A point is an ordered pair (x, y) of real numbers.

Not only can we think of these points (every point on the plane) as number pairs, but we can also handle them mathematically as number pairs. They are no longer geometrical points; they are things of arithmetic. This new approach works both ways, for we can also think of any pair of numbers as a unique point on the plane. The briefest glance tells us in which quadrant it belongs, the slightest effort places it exactly, no longer a number pair but a point, again a thing of geometry.

A point, we say then, is an ordered pair of real numbers (x, y). When we know the values for x and y, we know where the point lies on the plane.

But what happens to the values of x and y when our point moves about the plane leaving in its path a trail which we call a line?

If we take the point $(-3, -3)$, which we located on the diagram above, and proceed to move it so that its path is a straight line toward and through O, the origin, we find that the values of x and y change continuously. But one thing about them does not change: the x coordinate remains always the same as the y coordinate, just as -3 is the same as -3. If we move the point so that its path is a straight vertical line, the x coordinate remains the same (-3), but the y coordinate changes constantly. Moving the point on a horizontal line, we find that the reverse is true: the x coordinate changes constantly while the y coordinate remains the same (-3). We have here three

70

distinct lines, all passing through the point (—3, —3).
When we further consider the number pair (—3, —3), we
are immediately aware that the x and y coordinates add
up to —6. What happens when we mark on the plane near
it the other points the coordinates of which also add up to
—6? We find that we are mapping yet another straight
line which also passes through the point (—3, —3):

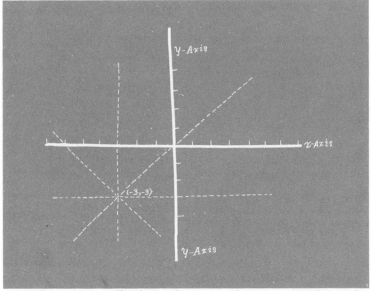

Like points, all of the lines we have mapped can be
uniquely identified in the terms of their x and y coordinates:

$$x = y$$
$$x = -3$$
$$y = -3$$
$$x + y = -6$$

But—we may object—we call these "paths of points"
or "lines" in geometry, but aren't they just equations in
algebra? Quite right. In analytic geometry we find that just
as points on the plane can always be expressed as ordered

pairs of real numbers (x, y), straight lines can always be expressed as equations of the first degree in two unknowns:

$$ax + by + c = 0^*$$

Although we have not done so above, we can express all of the equations we have listed in this standard form. The last line we graphed, for example, has the equation

$$x + y + 6 = 0$$

Since we have geometrized algebra at the same time we have algebrized geometry, such equations are now known as *linear* equations.

We have seen how the point of Euclid's geometry has become an ordered pair of real numbers and the line, the graph of an equation in the first degree with two unknowns. Now we must see what has become of Euclid's circle in this new number-based geometry.

On the Cartesian plane we draw a circle of unit radius with its center at the origin:

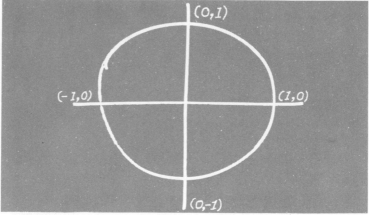

* It was Descartes who originated this form of expressing any equation so that the right-hand side is 0. He also started the custom of using the last letters of the alphabet for the unknowns and the first for the constants.

What can we say about the points on the line, the circumference of the circle, which will identify it algebraically as we have already identified straight lines? We see that the four points of the circumference which fall upon the x and y axes can be easily identified as $(0, 1)$, $(1, 0)$, $(0, -1)$ and $(-1, 0)$. What have these four pairs of numbers in common which might enable us to formulate a general rule for finding any point on the circumference of the circle we have drawn? Only that in each case the x and y coordinates add up to ± 1. Is this the general rule we are looking for? No, for when we locate the point $(\frac{1}{2}, \frac{1}{2})$, the coordinates of which also add up to 1, we find that this point falls inside the circle and not on the circumference. We are not so far off, though, as we appear to be.

The golden thread of the theorem of Pythagoras runs through our new algebrized geometry. If we draw a right triangle on the plane using the radius of our circle as the hypotenuse of the triangle, we can see by the Pythagorean theorem that r^2, or in this case 1^2, must be equal to the sum of the squares of the other two sides:

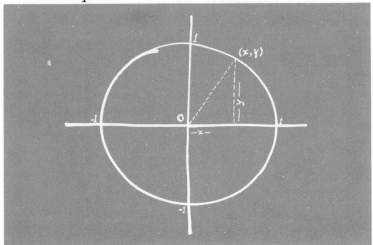

In the case of this particular triangle, the two sides are equal to the x and y coordinates of the point where the hypotenuse of the triangle cuts the circumference of the circle. We can say, therefore, that

$$r^2 = x^2 + y^2$$

or, for this particular circle:

$$1^2 = x^2 + y^2$$

Every point on the circumference of the circle we have drawn must be a number pair such that the sum of the squares of the two coordinates is 1:

$$\left(\frac{1}{4}, \frac{\sqrt{15}}{4}\right), \left(\frac{1}{3}, \frac{\sqrt{8}}{3}\right), \left(\frac{1}{2}, \frac{\sqrt{3}}{2}\right), \ldots$$

The way in which we have obtained this formula is in one sense a long way from the Greeks; yet in another sense it is as old as geometry itself. Regardless of where our circle lies on the Cartesian plane, we can by means of the Pythagorean theorem, with only a simple variation on the method above, express it as a similar equation of the second degree in two unknowns. This "distance formula" of analytic geometry, which is just as valid in three dimensions as in two, in seventeen as in three, runs concurrently with the geometry of Descartes as it is extended from two dimensions to three to four. . . . But that is a story for another chapter. In the meantime we can see that just as the equation of the circle is derived by means of the distance formula, so can the equation for any and all of the conic sections. For they, like the circle, depend essentially on a distance ratio. It can be proved that the curve of any conic is the graph of an equation of the second degree. It can also be proved that, conversely, any

74

curve defined by an equation of the second degree

$$ax^2 + bxy + cy^2 + dx + ey + f = 0$$

is one of the conics.

Although we have gained only the threshold of the new world of curves and their equations which was opened up by the invention of analytic geometry, let us turn back now to the Delian problem, which started us on our journey. The oracle had advised doubling the cubical altar of Apollo to appease the gods. The Greeks had assumed that the construction must be made only by the instruments of pure geometry; but, failing to solve the problem with straightedge and compass alone, they had toyed with certain other solutions mainly effected by devices for drawing one or more of the conic sections. We recall that they could not even state the problem with the vividness and suggestiveness which algebraic notation gives us, nor could they express the relatively simple characteristic properties of the sections with any degree of economy. Paragraphs of cumbersome technical vocabulary led to the enunciation of truths which can be expressed today with half a dozen letters, plus and minus, and the equals sign. And yet—Menaechmus solved the problem of doubling the cube by the intersection of conic sections!

Let us look at the solution of this same problem, using the powerful new tool which Descartes put in the hands of the mathematicians of the Renaissance when he invented analytic geometry.

We first graph the equation

$$x^2 = y$$

which is an equation for a parabola, and then

$$xy = 2$$

which is an equation for a hyperbola:

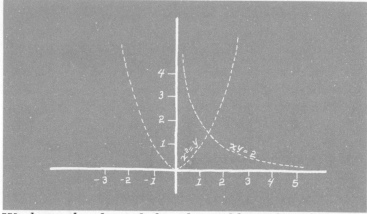

We have already said that the problem of doubling the unit cube is in modern notation the problem of solving for x the equation

$$x^3 = 2$$

Now if we consider the coordinates of the point at which the two curves graphed above intersect, we shall see that by the formula for the hyperbola the product of the x and y coordinates at this point must be 2; but by the formula for the parabola the y coordinate must also be equal to the square of the x coordinate. If we substitute the x^2 value for y (in the first equation) for the y in the second equation, we obtain

$$x^2 \cdot x = 2$$

or

$$x^3 = 2$$

The point of intersection of these two curves is then the solution of the Delian problem. If we take the length from the origin to the x coordinate of the point of intersection,

76

we shall have the necessary length for the side of our new cube, which will be twice the volume of the unit cube!

The Delian oracle is long silent, the original altar is in dust, the plague has been replaced by a thousand other plagues. To solve a problem made difficult by purely arbitrary restrictions, the conic sections have been discovered, analytic geometry has been invented. We are able to illustrate in this book the length of side of a cube which will be twice the size of the unit cube. But still the gods would not be satisfied, for the size of the new altar must be determined *by straight line and circle alone!*

FOR THE READER

The reader may enjoy graphing the following equations, the first few points of which are already indicated:

$$x = y \ (1,1), (2,2), \ldots$$

$$x - y = 2 \ (2,0), (3,1), \ldots$$

$$4x + 3y = 18 \ (0,6), (1, 1\tfrac{4}{3}), \ldots$$

$$y = x^2 - 2 \ (0,-2), (1,-1), \ldots$$

$$xy = 4 \ (1,4), (-\tfrac{1}{2},-8), \ldots$$

$$x^2 + y^2 = 4 \ (0, \pm 2), (-1, \pm\sqrt{3}), \ldots$$

6

How Big?
How Steep?
How Fast?

How big? how steep? how fast?

Among these three apparently unrelated questions there exists a deep and unexpected point of contact which can serve us as an introduction to the calculus, one of the most powerful tools of mathematics. It is by means of the calculus that mathematics has been able to make an effective attack on those problems which in earlier times admitted only of approximate answers.

Invented in the seventeenth century by Sir Isaac Newton (1642-1727) and Gottfried Wilhelm von Leibniz (1646-1716), who worked independently, the calculus had had its beginnings long before in two purely geometrical problems: how to compute an area bounded by a curve and how to draw a tangent to a curve at any point. More than nineteen hundred years before either of the inventors of the calculus was born, these two problems were solved (for special types of curves) by Archimedes (b.c. 287?-212), who used what were essentially the methods of the calculus of Newton and Leibniz.

How big?

The question *How big?* was one of the first to which mathematics sought an answer, and one of the first to which it found one, although not for all cases.

Given a rectangular area, it is a simple matter to compute that area as a sum of unit squares or, as we more often express it, the product of length and width. Given a triangular area, it can be shown that the area of a triangle is half that of a rectangle with the same base (width) and height (length) or, again, as we more often express it, one-half the product of base and height. Since any straight-edged surface, no matter how irregular its boundary, can be subdivided into triangles, the only remaining problem is to find the area bounded in part or in whole by a curve.

One method of doing so is to divide the area insofar as possible into rectangles and add together the areas of these. In the first figure following, it is clear that the sum of the areas of the rectangles, which we can compute exactly, gives us a fair approximation. In the second, it is

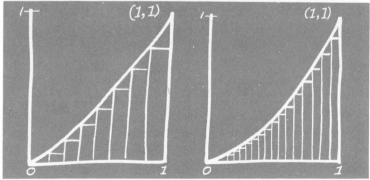

clear that more rectangles give an even more accurate approximation of the area which lies under the curve. We can continue, indefinitely, dividing the area into more and more rectangles and including as a result more and more of the total area under the curve. When we say that we can continue indefinitely, this is just what we mean: there is no limit to the number of rectangles into which we can divide the area—the number can "approach infinity." There is, however, a very real limit to the sum of the areas, no matter how many rectangles we use: *for the sum can never exceed the area under the curve.*

This limit provides us with a mathematically precise definition of what we mean by the area under the curve. It is the limiting value of the sum of the areas of the rectangles as the number of rectangles becomes indefinitely large.

Is this a satisfactorily accurate method of determining area? It is indeed. How very accurate it is can best be seen by applying it, not to a curved figure, the exact area of which we do not already know, but to a straight-edged figure like a triangle, the area of which we know is one-half the product of base and height. By this formula the

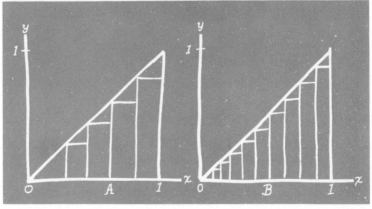

area of the triangle opposite is exactly ½. Since the y co-ordinate of any point on the hypotenuse has the same value as the x coordinate, we can easily determine the dimensions of each rectangle. In Fig. A, where we have divided the triangle into five intervals, the width of each being ⅕ of the base, we get the following sum when we add the areas together:

$$\frac{1}{5} \cdot \frac{0}{5} + \frac{1}{5} \cdot \frac{1}{5} + \frac{1}{5} \cdot \frac{2}{5} + \frac{1}{5} \cdot \frac{3}{5} + \frac{1}{5} \cdot \frac{4}{5} = \frac{10}{25}$$

But in Fig. B, where we have divided the base into tenths, we get a sum which is closer to ½, the true area of the triangle:

$$\frac{1}{10} \cdot \frac{0}{10} + \frac{1}{10} \cdot \frac{1}{10} + \frac{1}{10} \cdot \frac{2}{10} + \ldots + \frac{1}{10} \cdot \frac{9}{10} = \frac{45}{100}$$

By increasing the number of intervals from 5 to 10, we have brought our approximation from .40 to .45. The area with 50 intervals would be .49; with 100 intervals, .495. If we take n as the number of rectangular intervals into which we divide the triangle, we obtain the following general formula for the sum of the areas of n rectangles:

$$\frac{1}{n} \left(\frac{0}{n} + \frac{1}{n} + \frac{2}{n} + \ldots + \frac{n-1}{n} \right) =$$

$$\frac{1}{n^2} \left(0 + 1 + 2 + \ldots + n - 1 \right) =$$

$$\frac{1}{n^2} \left[\frac{n(n-1)}{2} \right]$$

If the reader, using this formula, will compute the sum of the areas of five hundred and one thousand rectangular intervals, he will find that these and any other higher n he chooses to compute will yield sums *between* .495 and .50.

81

Under no circumstances will the sum of the rectangles into which he divides the triangle be more than .50. That this is true is intuitively clear when we look at the triangle being subdivided and note the tiny triangles above the tops of the rectangles which can never be included in the sum of the areas. It is also clear when we further simplify our general formula for the sum of the areas:

$$\frac{1}{n^2}\left[\frac{n(n-1)}{2}\right] =$$

$$\frac{1}{2}\left(\frac{n-1}{n}\right) =$$

$$\frac{1}{2}\left(1-\frac{1}{n}\right)$$

As n gets larger (i.e., we cut our triangle into more and finer rectangles), $1/n$ gets smaller. As this happens, the value of

$$\frac{1}{2}\left(1-\frac{1}{n}\right)$$

will approach ½, the actual area.*

This method of determining area was called by Archimedes the method of "exhaustion" and by Newton and Leibniz, "integration." The latter two were fortunate in having at their disposal a tool which was not available to Archimedes. This was the analytic geometry of Descartes, with which—as has been frequently pointed out—a moderately intelligent boy of seventeen can solve problems which baffled the greatest of the Greeks. This statement is made, not to discredit Archimedes, whose place with Newton and Gauss in the pantheon of mathematics is uni-

* We can achieve the same result by circumscribing our rectangles so that they include more than the area of the triangle. As the number of rectangles gets larger, the sum will approach, from above, the limit which is the area of the triangle.

versally acknowledged, but only to emphasize the power of the method of analytic geometry.

When we can place our curves and figures on the plane formed by the x and y axes, we have a great advantage over Archimedes. Curves, as we have already seen, are no longer merely beautiful lines but definite relationships among numbers which can be expressed in a most general form—for the whole extent of the curve—by algebraic formulas. The straight line, or "curve," which forms the hypotenuse of the right triangle on the lower part of page 80 is determined by the algebraic equation $y = x$. When we say this, we mean that the numerical value of the y coordinate at any point on the curve is the same as the numerical value of the x coordinate at that point. If we are given $x = 9$ at a given point, we know that $y = 9$; if $x = 21$, $y = 21$; and so on. The curve on the lower part of page 79 is determined by the equation $y = x^2$. On this curve the numerical value of the y coordinate is always the square of the value of the x coordinate: if $x = 3$, $y = 9$; if $x = 9$, $y = 81$; and so on. The reader will recognize the equation for the parabola we used to solve the Delian problem in Chapter 5.

This method of analytic geometry is even more useful in answering our second question than it was in answering the first.

How steep?

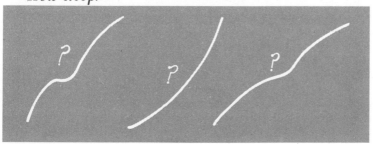

The question *How steep?*, like the question *How big?*, is simple enough to answer when only straight lines like $y = x$ are involved. If we look at the line below, we see that one measure of its steepness is the angle it makes with the x-axis and another is the ratio between the two coordinates x and y. If we take y/x as a measure of steepness, we see from the second figure that the greater y is in proportion to x, the steeper the line.

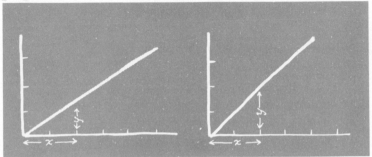

Neither method appears to be available to us when we want to determine the steepness of the parabola, or the curve represented by the equation $y = x^2$. Yet if we could draw a line which would have the same slope as the curve at some particular point, the same two methods of measuring steepness would serve.

Although the problem of determining such a line was solved by Archimedes in the special case of the spiral, it was not solved generally for all curves until, in the century before Newton and Leibniz, Fermat developed a general method of drawing a line (called a tangent) which touches a curve at only one point and hence has the same slope as the curve at that point.

When our curve is the arc of a circle, a line erected perpendicular to the radius at the point where it cuts the circumference will be tangent to the circle at that point. If we place the circle on the Cartesian plane with its cen-

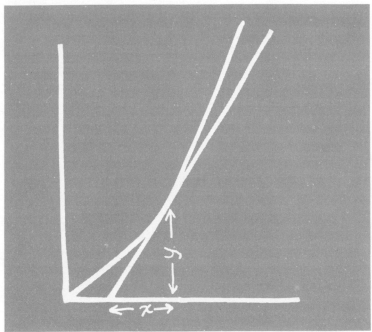

ter at the origin, the lines constructed perpendicular to the y-axis at the points where it cuts the circumference will be parallel to the x-axis and will represent the highest and lowest points (or extrema) of the curve. The determination of such high and low points for any curve was

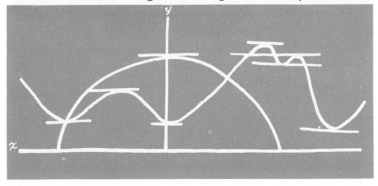

the particular problem which interested Fermat and for which he created a general method for drawing tangents.

To draw a line tangent to the point *P* in the figure below, according to Fermat's method, we mark on the curve in the neighborhood of *P* another point *Q* and draw a line from *P* to *Q*. As we slide the point on the line now marked *Q* along the curve toward *P*, always keeping the line *PQ* going through *P*, the closer *Q* gets to *P*, the more nearly will the line *PQ* represent the slope of the curve at *P*. In the language of the calculus, as *Q* is allowed to approach *P*, the line *PQ* will approach *a limiting position* which is the desired tangent to the curve at *P*.

These two geometrical problems, computing the area bounded by a curve and finding the slope of a curve at a given point, are at the very foundations of the calculus. The first is the fundamental problem of the *integral* calculus; the second, of the *differential* calculus. Both, as we have seen, were recognized from antiquity, tackled and partially solved long before the invention of the calculus in the seventeenth century. Newton and Leibniz were the first to recognize that these two problems were but facets of one and the same problem, and that the integral and the differential calculus were essentially one—*the* calculus. The theorem which states this truly deep relationship was discovered independently by both of them. It is the Fundamental Theorem of the Calculus.

Although the theorem cannot be stated or understood without some grasp of the technicalities of the calculus, the glimpse it can give us of this mighty tool in action is well worth the effort required to follow unfamiliar symbols and concepts. Already we have gained some idea of the two main concepts, those of limit and of function. These are basic to much of mathematics .beyond the calculus, and mathematicians can (and must) go on for pages defining precisely what they mean by *limit* and *function*. We, however, can make do with very little of this. We have seen that the area under the curve is defined as a limiting sum and the tangent to the curve as a limiting position. These give us an intuitive, if not too precise, idea of a limit. We have dealt with the curves of two functions so far, although we have never referred to them as functions. For our purposes, the simplest and most easily grasped definition of a function is a strictly mathematical one. A function is a rule *by which y is determined as soon as x is given.* If we apply this definition to the straight line determined by the equation $y = x$ and to the curve determined by $y = x^2$, we have no trouble in recognizing that both of these equations identify functions.

To express this concept of function there is a very simple and useful notation, $f(x)$, which is read "f of x" or "function of x." In the first of the examples we have given, $f(x) = x$; in the second, $f(x) = x^2$. Since any curve represents a value y determined by a value x at each point of the curve, we can identify any curve in a general way

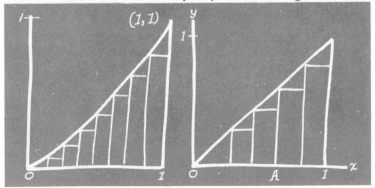

as $f(x)$, or as a function of x, even though we may not know the particular $f(x)$ that determines the curve.

Sometimes we are concerned not with the curve as a whole but with a particular point on the curve. Knowing that the x coordinate of the point is, say, 2, we can then write of y that $y = f(2)$. Whether y necessarily equals 2 depends solely upon the particular $f(x)$ which determines the curve as a whole. When the curve is determined by $f(x) = x$, $y = f(2) = 2$; but when the $f(x)$ of the curve is $f(x) = x^2$, then $y = f(2) = 4$.

Unfortunately, without understanding this much of the notion of function, we cannot possibly follow even the simplest applications of the calculus. At the end of this chapter, therefore, are a few problems which will enable the interested reader to clarify and make firm his own understanding.

With such a general notion of limit and function, we now need an understanding of the concept of *an increment* if we are to follow the Fundamental Theorem of the Calculus. The technique of the calculus depends essentially upon this concept. An increment is an arbitrarily small increase in x of $f(x)$ which, since $y = f(x)$, results in a corresponding (though not necessarily the same) arbitrarily small increase in the value of y. We symbolize the increment added to x by $\triangle x$ and the corresponding increment in y by $\triangle y$, and write

$$y + \triangle y = f(x + \triangle x)$$

where $\triangle$ is read "delta." To express what we have done in this general way, we do not have to know what x is, what the arbitrarily small increase in x is, what $f(x)$ is, or what the corresponding small increase in y is. We can even proceed, still not knowing the value of any of our terms, to express $\triangle y$, or the increase in y, solely in the terms of x.

$$y = f(x)$$
$$y + \triangle y = f(x + \triangle x)$$
$$\triangle y = f(x + \triangle x) - f(x)$$

Once we have expressed $\triangle y$ in terms of x, we can express the ratio $\triangle y / \triangle x$ in terms of x.

$$\frac{\triangle y}{\triangle x} = \frac{f(x + \triangle x) - f(x)}{\triangle x}$$

Perhaps we appear to be getting nowhere fast?

But it is one of the marvels of mathematics that such apparently pointless manipulation of symbols should be the source of the power of the calculus, one of the most practical of the many tools with which mathematics has outfitted modern science! Appearances to the contrary, we are getting somewhere—but fast. To see that we are, let us return to the curve of the parabola, which is represented by the equation $y = x^2$. We learned earlier how to determine the slope of such a curve at any given point, but now let us consider a less geometrical and more general question. What is the rate of change represented by this curve? How fast is y changing with respect to x? Actually, although these two questions sound quite different, they are the same as the question *How steep?*

Since in the case of this curve, $f(x) = x^2$, we know that the value of y is increasing as the square of the value of x:

x	0	1	2	3	4	5	6	7...
y	0	1	4	9	16	25	36	49...

Obviously y is increasing much faster than x. Between 0 and 1, both x and y increased by 1; but between 6 and 7,

89

x still increased by only 1 but y increased by 13. Between 0 and 7, x has gained 7 points while y has gained 49. The average gain of y in proportion to that of x is 7 to 1. But how *fast* is y gaining on x?

Let us apply the method of the calculus to this problem: a method which appeared a few pages back as a meaningless manipulation of symbols. We begin by adding an arbitrarily small amount to x in $f(x)$ so that we have instead of $f(x)$, $f(x + \triangle x)$. Since $y = f(x)$, the new value of y is $y + \triangle y = f(x + \triangle x)$. Now let us substitute for $f(x)$ in its general form the specific function x^2 with which we are dealing. We begin with

$$y = x^2$$

After we add the increment to x^2, we have

$$y + \triangle y = (x + \triangle x)^2$$

When we express $\triangle y$ in terms of x, we get

$$\triangle y = (x + \triangle x)^2 - x^2 = x^2 + 2x \cdot \triangle x + (\triangle x)^2 - x^2$$
$$= 2x \cdot \triangle x + (\triangle x)^2$$

If we now express the ratio between $\triangle y$ and $\triangle x$ in the terms of x and then cancel out identical terms in numerator and denominator, we arrive at

$$\frac{\triangle y}{\triangle x} = \frac{2x \cdot \triangle x + (\triangle x)^2}{\triangle x} = 2x + \triangle x$$

Recalling that when we first added $\triangle x$ to x in $f(x)$, we defined it as "an arbitrarily small increase," we realize that as we choose smaller and smaller amounts for $\triangle x$, i.e., $\triangle x$ approaches 0, the limiting value of the ratio $\triangle y / \triangle x$ will be $2x$. This is the rate of change of y with respect to x when $f(x) = x^2$.

We can see that $2x$ actually is the rate of change, or,

to express it in a different way, the slope of the curve at a given point. We plot the parabola and then at any point draw a line the slope of which is equal to twice the value of the x coordinate of the point. For instance, at $x = 1$ the slope should be 2; so we line up our straightedge with a point 1 unit over and 2 units up from our given point on the curve. The slope of the line we draw will then be 2,

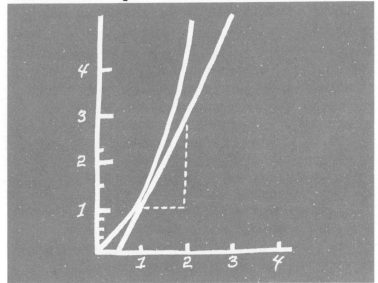

and we can see that this line does represent the slope (or rate of change) of the curve at this point.

How big? How steep? How fast? We have said that there is a fundamental point of contact among these three questions. We have shown that the answers to the last two are essentially the same. *How steep? = How fast?* Now we shall show the relation of the first to these two. That these three questions are so related has been called "one of the most astonishing things a mathematician ever discovered."

We begin by taking the area under a curve which we

91

can identify in a general way as $f(x)$. We have seen that a curve is a function of x since each x coordinate determines a y coordinate and hence the curve itself. The area under a curve is also a function of x but in a somewhat different sense. It is clear from the diagram below that if we take a as the x coordinate of the left-hand boundary of the area we wish to compute, and b as the x coordinate of the right-hand boundary, moving b to the right on the x-axis will increase the area. In this sense the area under a curve is a function of (i.e., is determined by) the value of the x coordinate at its right-hand boundary.

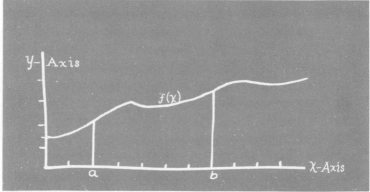

Since, although the area is also a function of x, it is not the same function as that which determines the curve above it, we represent the curve by $f(x)$ and the area by $F(x)$. This can be easily seen in the curves below. On the left we have a triangle under the curve $f(x) = x$, the value of each y coordinate being the same as that of the x coordinate of any point on the curve. If we compute the area of this triangle at each x coordinate as one-half of x^2 (or half the base times the height), we find that the curve representing the area as a function of x, or $F(x)$, is an entirely different curve, $F(x) = \dfrac{x^2}{2}$:

92

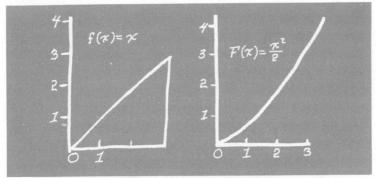

Now let us return to our main problem.

To determine the area under the curve between a and b we proceed in the by now somewhat familiar method of the calculus. We go a little farther to the right on the x-axis and add to x (represented on the diagram by b) an arbitrarily small distance which we call $\triangle x$. This results in an appropriately small increase in the area under the curve, which we call $\triangle A$.

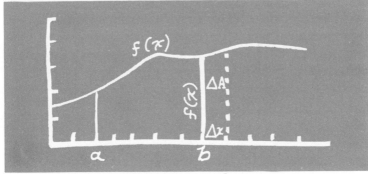

Instead of $A = F(x)$ we now have

$$A + \triangle A = F(x + \triangle x)$$

and by subtracting the original area from the enlarged area we can determine the value of $\triangle A$.

$$\triangle A = F(x + \triangle x) - F(x)$$

93

If we look at our diagram we can see by inspection that $\triangle A$, as well as having the value given above in terms of x, has also the *approximate* value of $\triangle x$ times $f(x)$, which would be the area of the largest rectangle we could inscribe in $\triangle A$. The ratio $\triangle A / \triangle x$ is then approximately $f(x)$.

$$\lim_{\triangle x \to 0} \frac{\triangle A}{\triangle x} = \lim_{\triangle x \to 0} \frac{\triangle x \cdot f(x)}{\triangle x} = f(x)$$

From the above we see that the area under the curve $f(x)$ is determined by a function $F(x)$ which has the property that its rate of change, or derivative as it is technically called, is $f(x)$!

Since $F(x)$ answers the question *How big?* and $f(x)$ answers *How steep?* and *How fast?*, we find all three inextricably bound together. This is the fundamental relationship of the calculus—"one of the most astonishing things a mathematician ever discovered"!

With a brief explanation of two notations which we have not already met, we are now ready to state and follow the Fundamental Theorem of the Calculus. For the derivative of $F(x)$, we shall use the notation $F'(x)$; and for the area under $f(x)$ between $x = a$ and $x = b$, the notation below.

$$\int_a^b f(x)\,dx$$

The Fundamental Theorem, discovered independently by Newton and Leibniz, states:

If $f(x)$ is continuous and $F'(x) = f(x)$, then

$$\int_a^b f(x)\,dx = F(b) - F(a)$$

94

Let us apply this formula to the area under the line $y = x$ between 0 and 1, which we know is ½, and the area under the curve $y = x^2$ between 0 and 1, which we do not know. In the first case we must have a function of x, the derivative (or rate of change) of which is x. Since we earlier determined the rate of change of x^2 as $2x$ (on page 90), we can surmise that the derivative of ½x^2 is x. In the second case, the reader may be interested in working out (as on the same page) that the derivative of ⅓x^3 is x^2.

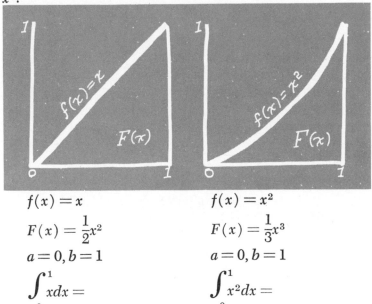

$f(x) = x$

$f(x) = x^2$

$F(x) = \frac{1}{2}x^2$

$F(x) = \frac{1}{3}x^3$

$a = 0, b = 1$

$a = 0, b = 1$

$$\int_0^1 xdx =$$

$$\int_0^1 x^2 dx =$$

$\frac{1}{2} \cdot 1^2 - \frac{1}{2} \cdot 0^2 = \frac{1}{2}$

$\frac{1}{3} \cdot 1^3 - \frac{1}{3} \cdot 0^3 = \frac{1}{3}$

In the case of the triangle we know that the area is indeed ½, which the Fundamental Theorem gives us as the limit. In the case of the area under the parabola, we did not know but *now* we know that the area under the curve, defined as the limit, is ⅓.

95

Thus the Fundamental Theorem of the Calculus brings together the answers to the three questions we asked about curves and the areas which lie under them. *How steep?* has the same answer as *How fast?*, and the answer to *How big?* is the inverse of the other two.

It was because they perceived this underlying unity that Newton and Leibniz, who were by no means the first to use the methods of the calculus, are given the full credit for its invention.

FOR THE READER

The test below will enable the reader to make sure that he has a clear, if simple, notion of a function.

1. If $f(x) = x$, solve $y = f(5)$ for y.
2. If $f(x) = x^2$, solve $y = f(5)$ for y.
3. If $f(x) = x^2$, what are the y coordinates for $x = 1, 2, 3$?
4. If $f(x) = x$, what are the y coordinates for $x = 4, 5, 6$?
5. If $f(x) = 2x$, solve $y = f(8)$ for y.
6. If $f(x) = 2x - 1$, solve $y = f(3)$ for y.
7. If $f(x) = 1/x$, what is the value for y when $x = 7$?
8. If $f(x) = 1 - x$, what is the value for y when $x = 1$?
9. If $f(x) = x^3$, solve $y = f(2)$ for y.
10. If $f(x) = x + 3$, what is the value of y for $x = 7$?

ANSWERS

1. $y = 5$; 2. $y = 25$; 3. $y = 1, 4, 9$; 4. $y = 4, 5, 6$; 5. $y = 16$; 6. $y = 5$; 7. $y = \frac{1}{7}$; 8. $y = 0$; 9. $y = 8$; 10. $y = 10$.

7

*How Many
Numbers
Are Enough?*

1, 2, 3, . . . are enough numbers to count the objects before us; yet when we encounter the ordinary operations of arithmetic for the first time, we find that they are not nearly enough. We can subtract only when the number being subtracted is smaller than the number it is subtracted from; we can divide only when the number being divided is a multiple of the number being divided into it; we can extract a square root only when the number from which we are extracting it is a perfect square. For these simple operations to be always possible, we must have more numbers than 1, 2, 3, ...

We have seen in an earlier chapter how the necessity for a number for every point on the line resulted in the development of the concept of the arithmetic continuum. Now we shall see how the necessity for an "answer" to every problem in arithmetic resulted in a parallel development that went one step beyond the concept of a unique number for every point on the line—to a unique number for every point on the plane!

The necessary extension was a matter of centuries. Although we shall follow it in a more or less logical order, it was neither orderly nor logical.

Numbers began as a way of count-

97

ing. It seemed natural that a number should correspond to each thing counted, so in later times when all sorts of curious quantities were being used as if they were numbers, these original counting numbers came to be thought of as the *natural* numbers. "God made the integers," thundered a mathematician of the nineteenth century.* "All else is the work of man."

Numbers other than the original natural numbers turned up in the process of solving problems: first probably in making accurate measurements, later in finding the roots to equations. Even mathematicians had curious attitudes toward them. The Greek mathematicians used rational quantities, or fractions, but refused to call them numbers; and the beautiful theory of numbers which they created deals, to this day, only with whole numbers. The Indian mathematicians did not consider the negative solutions to equations as solutions "because people do not approve of negative roots." The mathematicians of the Renaissance, who solved otherwise unsolvable equations by acting as if -1 had a square root, uneasily dismissed $\sqrt{-1}$ (after they had used it) as "imaginary."

But in spite of the fact that the mathematicians did not really believe that anything other than 1, 2, 3, . . . was a number, they ended up by justifying their use of these quantities as numbers by the fact that they used them in the same way they used 1, 2, 3, . . . , adding, subtracting, multiplying and dividing them according to what they considered the *natural* laws of arithmetic.

Although very few of us could state these Laws of Arithmetic on a quiz program, we obey them almost unthinkingly. The Associative Laws, the Commutative Laws and the Distributive Law, as they are called, are no more than the formal statements of how the natural numbers

* Leopold Kronecker (1823–1891).

98

behave under the operations of addition and multiplication and, by implication, subtraction and division.

The Associative Law of Addition, for instance, tell us that when adding 1 and 2 and 3, we can perform the operation in several different ways and still get the same answer; and the Associative Law of Multiplication tells us the same thing in regard to multiplying:

$$1 + 2 + 3 = 6 \qquad\qquad 1 \times 2 \times 3 = 6$$
$$(1 + 2) + 3, \text{ or } 3 + 3 = 6 \qquad (1 \times 2) \times 3, \text{ or } 2 \times 3 = 6$$
$$1 + (2 + 3), \text{ or } 1 + 5 = 6 \qquad 1 \times (2 \times 3), \text{ or } 1 \times 6 = 6$$

It is important to note that the Associative Laws do not tell us that we can change the order of 1, 2 and 3 when we add or multiply them and still get the same sum or product. That is reserved for the Commutative Laws.

We are all familiar with the fact that if we take two of something like an apple and then three, we shall have as many apples as the person who first took three and then two. If we take two apples three different times, we shall have as many apples as the person who reached for the bowl only twice but took his apples three at a time. These simple facts of social life are formalized in the Laws of Arithmetic to the effect that addition and multiplication of the natural numbers are commutative operations:

$$2 + 3 = 3 + 2 \qquad \text{and} \qquad 2 \times 3 = 3 \times 2$$

The Distributive Law merely brings addition and multiplication together with the statement that $2 \times (1 + 3)$ is the same as $(2 \times 1) + (2 \times 3)$.

In the past, mathematicians firmly believed that these laws were as "natural" and God-given as the numbers to which they applied; yet all around them were "multiplications" and "additions" not associative or commutative.

99

We have actually seen that addition and multiplication, when applied to apples, are commutative; but do we know that they are always commutative in respect to things other than apples? We, like the mathematicians of the past, probably think that we do; but let us look for a moment at baseball hits instead of apples. If our team gets a three-bagger and a home run, the total number of bases hit will be the same whether we add

$$3B + 4B$$

or

$$4B + 3B$$

but there will be a considerable difference in the score depending on which hit was made first:

$$4B + 3B = 1 \text{ run}$$

but

$$3B + 4B = 2 \text{ runs}$$

If we buy an insurance policy after we have had an automobile accident, the result of the combination of accident and policy is quite different from what it would have been if the combination had been made in the reverse order:

$$\text{Policy} + \text{Accident} = \$1000$$

but

$$\text{Accident} + \text{Policy} = \$0$$

There are many other examples in everyday life where the order of combination changes the result of an operation. We offer these only to show that while it may be impossible for us to think of 2×3 as not being equal to 3×2, we can think of ab, under certain conditions, as not being equal to ba.

Subtraction and division are neither associative nor

commutative. But do multiplication and addition have to be associative and commutative? Does addition have to be distributive with respect to multiplication whenever we are dealing with quantities we choose to call "numbers"? Up until a little more than a hundred years ago, it was thought by all mathematicians that they did; that, in fact, they must. The Laws of Arithmetic were considered a logical necessity of number.

We have heard a great mathematician say that God made the integers and all else was the work of man. This was the attitude of mathematicians from the time of Pythagoras. To facilitate measurements and the solutions of equations, mathematicians might have to extend the concept of number to include quantities other than the integers, but they could at least see that, like the integers which God made, these followed the God-given Laws of Arithmetic. In all the extensions of the number concept which we shall describe in this chapter, this principle was followed. It was called the Principle of Permanence of Form; and it meant that the fundamental Laws of Arithmetic, which we have already examined, remained in force with the new numbers as well as with the old. This made everybody feel much better about using the strange new "numbers."

To understand the extensions which were made, we shall begin with a picture of the natural numbers marked off, unit by unit, upon a straight line extending indefinitely to our right:

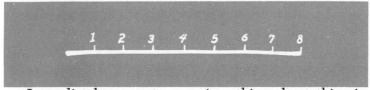

Immediately we note a curious thing about this pic-

101

ture. While 1 marks the distance 1 unit from the beginning of the line; 2, the distance 2 units from the beginning; 3, 3 units, and so on—there is no number among the original natural numbers which can mark the beginning of the line. Yet if we take away, or subtract, 4 units from the point marked 4, this beginning point is exactly where we obtain our answer. What is the answer to the question *How many is 4 — 4?* The answer is none at all or, numerically speaking, 0. So let us call 0 a number, since it answers the question *How many?* just as the other counting numbers do, and then let us mark the beginning of the number line with 0.*

Zero makes possible the subtraction of a number from itself.

But even with 0, subtraction is not always possible. We still cannot subtract a larger number from a smaller and get as our answer a number on the line above. When we take 6 from 5, we find that we are 1 unit short. In other words, we could perform the operation if we had one more unit to the left of 0. So, arbitrarily, we add it and an infinite number of such units. We extend the

* This is not at all the way that 0 was invented. It was invented, not as a number, but as a symbol to mark those columns in the representation of a number which contained no digits. The use of 0 made possible the representation of all numbers with only ten different symbols and was probably one of the most important practical inventions in the history of the world. The idea of 0 as a number (rather than merely a symbol) is not very important to anybody but a mathematician, to whom it is quite important. In the modern theory of numbers, 0 is usually treated as one of the *natural* numbers.

102

number line to the left of 0, and we mark it off in units just as we did the line to the right. Since these units are less than 0, we place a minus sign in front of them and call them *negative*. To be consistent, we must then place plus signs in front of what were once *all* the numbers, and call them *positive*. Zero has neither plus nor minus in front of it—is neither positive nor negative. The extended number line now looks like this:

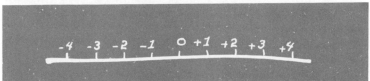

The negative numbers make subtraction always possible.

Now we come to division and face to face with the unpleasant fact that most divisions do not come out even. If we are to perform the operation of division whenever it is indicated and get an answer among the numbers on our line, we must have parts of numbers, divide our units into sub-units, and allow these to be "answers" too. Unless we do so, we can divide a number only into a multiple of itself.

Although we shall indicate on our extended line just those sub-units obtained by dividing the unit in half and then in half again, we must understand that to make division always possible we have to include among our new numbers every quantity which can be represented by the ratio of two whole numbers. We give these new quantities the Greek name of the *rational* numbers. As a class, the rationals include the whole numbers, for these can always be expressed as the ratio of themselves over 1. With the extension of the number concept to include fractional parts of the unit, our line begins to look like this on the portion between —1 and +1:

The rational numbers make division, except division by 0, always possible.

Things are getting a little crowded even with only the few numbers we are indicating on the number line. It is now, in fact, what mathematicians characterize as *dense*, which means that between any two numbers there is always another number. As we have seen, the Greeks at the time of Pythagoras, with far fewer numbers (for they had not extended their concept of number to include either 0 or the negative integers), thought that they had quite enough for all practical purposes, including the measurement of the Universe. We have also seen how the most shattering discovery in the history of mathematics was the discovery that this beautiful array of whole numbers and their ratios was not enough to furnish an exact measurement of the diagonal of the unit square. The square root of 2 was a non-rational number.

How many such non-rational, or irrational, numbers are there? Merely an infinite number—and this in spite of the fact that, as we have seen, the rational numbers are dense upon the line. By multiplying by itself a rational number which is not a whole number, we can never get a whole number as our result. All numbers, therefore, which are not perfect squares, or generally perfect powers, of some other whole number must have as their roots irrational numbers.

104

The irrational numbers make the extraction of roots of positive numbers always possible.

Up to this point we have been writing of these successive extensions of the concept of number as if they were things that we would be unable to live without. Yet the majority of people in even the most civilized countries do not consider 0 a number, but rather a symbol which is indispensable for the representation of numbers in the decimal system. It is most unusual to see the digits arranged in their natural order 0, 1, 2, 3, . . . , 0 usually being placed instead *after* 9. Only of late, with the "countdown," has 0 been publicly recognized as a number, and then it is counted back to, rather than up from. Although we are all familiar with debts, losses, arrears and such unpleasant figures, we never put a minus sign in front of them in our accounts, but write them in red. We treat both profit and loss as positive quantities and subtract the smaller from the larger to find out whether we are ahead or behind, and how much. We would find it difficult to live without the rational numbers, since sub-units of the unit are necessary for even fairly approximate measurements; but considering the infinities of rational numbers—infinity upon infinity—which are at our disposal, we use practically none of them. The ordinary foot ruler distinguishes only to $\frac{1}{16}$ of an inch. Since we can place any irrational root to as many decimal points as we wish, and have the time and energy to compute, it is obviously of no great concern that we cannot place it exactly.

The truth of the matter is that the successive extensions of the number system took place, not to make the ordinary operations of arithmetic always possible in everyday life, but to make them always possible in algebra. If we are to be generally effective in the solution of algebraic equations, we must know before we start that there exists a number which will satisfy each respective unknown in an equation. This does not mean, even when we come now to the final extension to the so-called *imaginary* numbers, that the successive extensions of the number system have no practical value. Algebra is one of the most practical subjects in the world. Just ask any scientist!

But let us imagine for a moment that we are limited in our algebra to solutions for x which are among the original natural numbers. Then let us try to solve the following simple equations by finding in each case a value for x:

$$x + 3 = 3$$
$$x + 3 = 0$$
$$3x = 1$$
$$x^2 = 3$$

We can see by inspection that to solve these equations, we must in each case extend the concept of number from the natural numbers—to zero, to the negative numbers, to the rational numbers, to the irrational numbers.* Not one of these equations would have a solution if we were limited in our algebra to the original natural numbers! If this restriction had been placed on our solutions, we would have seen that we had to stop before we started. But in more complicated equations we cannot see so easily that there is no solution for x. If we are to proceed in our manipulations of the symbols with any assurance that these manipulations are not a waste of time, we must know before we start that in every case there exists a *number* for x.

---

* In actuality we have added many numbers which we do not need for the solution of algebraic equations—numbers, called "transcendental," which *cannot be roots of algebraic equations.* We shall hear more of these numbers later.

Having extended the number system four times already, we can now find roots for any of the equations above and for any similar but much more complicated equations. Yet, we are not through. There are still comparatively simple equations for which we can find no roots at all among the numbers we already have. Such an equation is

$$x^2 + 1 = 0$$

It is obvious that if we are to add 1 to x^2 and obtain 0, x^2 must have the value of —1. It is equally clear that x then must have the value of $\sqrt{-1}$. BUT, under the rules by which the negative numbers were allowed to be brought into our number system, it was implicitly stated that a negative number could not have a square root! Recall the rules for multiplying positive and negative numbers, which were necessary for maintaining the Principle of Permanence of Form. A positive number multiplied by a positive number yields a positive number, as does a negative number multiplied by a negative number:

$$(+2) \times (+2) = +4$$
$$(-2) \times (-2) = +4$$

Only when we multiply together a positive and a negative number do we get a negative product:

$$(-2) \times (+2) = -4$$
$$(+2) \times (-2) = -4$$

We must remember that $+2$ and -2 are two different numbers, located at two entirely different points on our number line. But by definition a square is the product of a number *multiplied by itself*. Under this definition a negative number simply cannot be a square. Yet there is our equation—

107

$$x^2 + 1 = 0$$

If we cannot find a solution for this equation, we shall be severely handicapped in our algebra. We shall have failed in our avowed purpose of extending the number system so as to make the operations of arithmetic always possible. We shall have to concede that any equation for which $x^2 = -1$ *has no root.*

Let us not give up too easily. In the first half of the sixteenth century Girolamo Cardano (1501-1576), saying frankly that roots of negative numbers were "impossible" —there could be no such roots!—nevertheless began to use in solving otherwise unsolvable equations a symbol which he called the square root of -1. Since he did not consider this symbol a real number (for he knew as well as anyone that there could not be a number which when multiplied by itself would produce -1) Cardano called his symbol an *imaginary* number. The strange thing was that by using such imaginary numbers when necessary, Cardano found that he could obtain very real, practical results with equations which otherwise he would not have been able to solve!

But let us return for a moment to our own extension of the concept of number. How can we, refusing to have anything to do with Cardano's highhanded invention of a "number" for the square root of -1, go about finding such an impossible root for an equation like $x^2 + 1 = 0$ in a logical and orderly extension of our concept of number? There is no root among the integers, the rationals or the irrationals. At this point we cannot change the rules under which we brought these quantities in as numbers. We cannot, for instance, say that a negative number multiplied by a negative number yields a negative number, for that would involve us in impossible contradictions. It was

108

to avoid the contradictions that we made the rules in the first place. There is only one thing we can do. Just like Cardano, we can make up another number. We can simply define it as $\sqrt{-1}$ and call it i (for Cardano's imaginary number).

We have no everyday justification for what we are doing. We can compare the negative numbers to things like debts and temperatures below zero and the years before the birth of Christ, but the number i we can compare to nothing in everyday life. It was for this reason that mathematicians, although they went right along using i to solve equations, felt a little guilty about what they were doing. God, they felt, had made the whole numbers. If He had wanted man to have them, He would have made negative numbers and given them square roots!

Yet the extension of the concept of number to the imaginary numbers parallels in a logical and orderly way the extension to the negative numbers. The negative numbers were invented to make subtraction always possible; the imaginary numbers were invented to make extraction of roots always possible. There was only one condition upon the admission of negative numbers to the number system: they must be used in accordance with the Laws of Arithmetic, the Principle of Permanence of Form must be maintained at all costs. This same condition was imposed upon the imaginary numbers. They were just as much numbers, and every bit as "real," as the negative numbers. Unfortunately, in the beginning they were called "imaginary" by Cardano and the name has stayed with them—and undoubtedly always will.

Today the words "real" and "imaginary" are used to distinguish the two axes of a number plane which is as real as the plane of analytic geometry, and identical with it. Obviously i and its multiples $2i$, $3i$, . . . , cannot go on our number line, since all the points on the line are already accounted for by numbers. They can, however, have a line of their own—the pure imaginary line which, like the y-axis of the Cartesian plane, is perpendicular to the real number line, or x-axis, at 0:

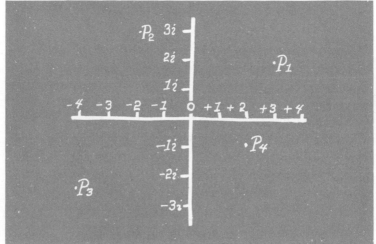

With this geometric interpretation, we find that our seemingly "imaginary" numbers begin to assume an everyday reality. Like the reals they have a line of their own. Combined with the reals, they serve to locate uniquely each point on the plane. These new combinations which do the same job as Descartes' pairs of real numbers (x, y) represent, however, an important advance in our concept of number. While Descartes' real-number coordinates are "pairs" of numbers, these combinations of real and imaginary coordinates are *individual numbers*.

These new numbers of the form $(x + yi)$ are called complex numbers because they have more than one part. They are represented abstractly as $(x + yi)$ where x and y are real numbers and i is defined as $\sqrt{-1}$. When x has the value of 0, the "complex" number becomes a pure

110

imaginary $(0 + yi = yi)$, while when y has a value of 0 it becomes a real number $(x + 0i = x)$. The pure imaginaries and the reals are, therefore, merely sub-classes of the complex numbers:

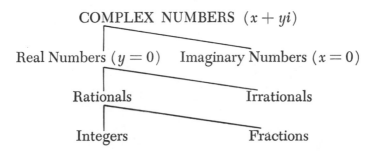

COMPLEX NUMBERS $(x + yi)$

Real Numbers $(y = 0)$ Imaginary Numbers $(x = 0)$

Rationals   Irrationals

Integers   Fractions

The imaginaries make the extraction of the roots of negative numbers always possible.

We have come a very long way in our extension of the concept of number. We began with the natural numbers, which could be paired in one-to-one correspondence with objects which were to be counted. By retaining the rules which we had made up for the behavior of these numbers, we were able to extend without logical difficulty our concept of number to the so-called "real" numbers, which could be paired in one-to-one correspondence with every point on a line. Still retaining the same rules, we further extended our concept of number to the complex numbers, half "real" and half "imaginary," which could be paired in one-to-one correspondence with every point on a plane. We have enlarged our number system, step by step, so that now for every operation of arithmetic we can obtain an answer within our number system. Just as with the original natural numbers we could *always* add or multiply, now with the complex numbers we can always add, subtract, multiply, divide and extract roots.

111

But we are still troubled.

We saw that the extensions to 0 and the negative numbers made subtraction always possible; the rationals made division always possible; the irrationals made the extraction of roots of positive numbers always possible; the pure imaginaries and the number i that generates them made the extraction of roots of negative numbers always possible.* We now have a number for every point on the real axis and every point on the pure imaginary axis and, also, a number for every point on the plane. Surely these should be enough numbers to make the operations of arithmetic always possible and to provide every algebraic equation with a root! *But what about an equation like this one?*

$$x^2 - i = 0$$

Won't we need to extend our number system once again, beyond i to $\sqrt{-i}$?

The answer to this question is a very simple one, which mathematics can offer with all the finality of mathematical proof. The answer is *no*. We have gone as far as we need to go. It can be shown—and this is known as the Fundamental Theorem of Algebra—*that any algebraic equation has a root within the system of complex numbers.*

To mathematicians i, the square root of —1, is the wonderful square root. In the satisfying language of mathematics it is both necessary and sufficient.

That pesky equation? Don't we need a square root of i to get a root for that x? Oh no,

$$x = \pm\left(\frac{1 + i}{\sqrt{2}}\right)$$

Multiply it out, and see for yourself!

* $\sqrt{-2} = i\sqrt{2}$, and so on.

One of the quickest ways to get rid of the idea that complex numbers are mere figments of our imagination is to pin them down geometrically. This is exactly what was done in the early nineteenth century when it was shown by Gauss and others that the domain of the complex numbers is mathematically equivalent to Cartesian plane geometry.

Geometrically, a complex number $(x + iy)$ is considered as a composition of the two vectors of its real and imaginary (or x and y) coordinates. In non-mathematical language we can think of it as the diagonal formed when we complete a rectangle from these coordinates:

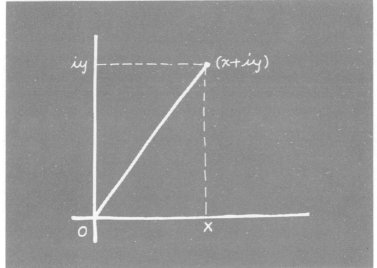

This same idea is extended to the geometrical definition of addition of two complex numbers (which are, of course, in themselves additions of real and imaginary parts). To add two complex numbers $(x + iy)$ and $(u + iv)$, we simply add real and imaginary parts sep-

113

arately and obtain as our answer the complex number

$$(x + u) + i(y + v)$$

Geometrically, we "complete the parallelogram" begun by the vectors of the two numbers, as below:

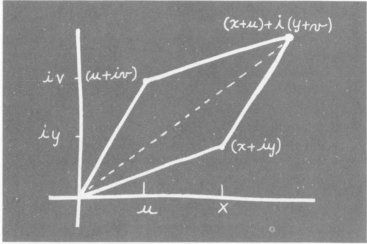

The reader will find it interesting to add the complex numbers below by both methods:

$$(3 + 2i) + (4 + 5i)$$
$$(2 + 5i) + (6 + 2i)$$

8

*Enchanted
Realm,
Where Thought
Is Double*

THE ART OF GEOMETRY. THE GEOMETRY
of art.

From the annexation of these two
territories mathematics gained what
has been called "an enchanted realm,
where thought is double and flows in
parallel streams." But such is the underlying unity of all mathematics that
just as we found the lines and circles
of Euclid and the conic sections of
Apollonius in the graphs of algebraic
equations on Descartes' numbered and
coordinated plane, so we meet them
again in this new domain.

The "enchanted realm" is projective geometry, the mathematics which
was born in the struggles of the early
Renaissance painters to transfer three
dimensions to two without losing the
appearance of reality.

The approach of projective geometry to the familiar subject matter of
geometry is *synthetic*—we proceed by
synthesis, or putting together, from
the figures to the principles. The approach of analytic geometry is, as its
name tells us, the direct opposite. We
proceed by analysis, or taking apart,
from principles to figures. Yet, curiously, the small volumes which formally introduced these two new geometries to mathematics were published
within a few years of one another!
Although neither book was to have

115

much influence on the immediate mathematics of its time, the very look of geometry had been irrevocably changed with their publication.

In the geometry of Euclid we drew circular circles and square squares, equilateral triangles with three equal sides, and parallel lines which never met. But these are not what we see. To the human eye, circles are not generally circular, squares are not square, equilateral triangles do not have equal sides or equal angles, and parallel lines approach one another. The only time we come even close to seeing these shapes in their pure Euclidean form is when we look at them head-on so that our eyes are more or less in line with the center of the figure. If we had just one eye, which could then be directly in line with the axis, we would be able to see them even more "accurately."

Given the problem of drawing a three-dimensional cube (or a box or a house) on a two-dimensional piece of canvas, we are immediately confronted by a paradox. We *know* that each of the six sides of the cube is a square—a quadrilateral with equal sides and equal angles —and that all six squares are the same size. If we look at the cube head-on, we see just one square; but if we draw the cube as a square, it certainly will not look to the eye like a cube. Our eyes tell us that the square is a cube because, having two eyes, we have brought together two slightly different views of the face of the cube and thus obtained at least a sense of depth. The square on our canvas is a Cyclops view. It has no depth unless the square face of the cube is seen among other three-dimensional objects which are not drawn head-on. To make the cube by itself appear solid to the eye, we must draw it from an angle which shows more than one face; and when we try to do just this, we find that not one of the faces which we draw is still a square!

We see now as we look at the cube from different angles that the faces change shape as we change our point of view. We may see one, two or three faces of the cube at once; but all will be different. We *know*, however, that they, as well as the faces which we cannot see, are all the same. After all, what is a cube but a solid with six square faces! Even though they may not look as if they are, all lines and all angles must be equal.

There is one thing about the faces of the cube which does not change. We may look at the cube from above, from below, from the left, from the right, the fact remains that every face we see is always a closed figure bounded by four straight lines—a quadrilateral of Euclid's geometry.

What rules determine these new quadrilaterals which are no longer squares? It was this question and similar ones about other geometrical shapes which led the painters of the early Renaissance to investigate and formulate the principles of perspective.

The word *perspective* in its original form means "to see through." It is an almost literal statement of what the

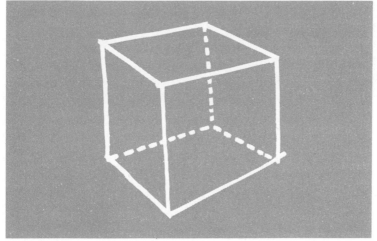

painter conceived himself to be doing. From every visible point of the object which he saw before him, a ray of light entered his eye. If a pane of glass were to be placed between the objects and his eye, each of these rays would pierce it at a definite point. The painter then could conceive of his painting as an imaginary glass through which he saw the scene. By drawing on his canvas the outline of objects exactly as they would appear on the imaginary glass transposed between his eye and the objects themselves, he could paint "what he saw." Some painters of the day actually used such mechanical aids.

Since these men of the early Renaissance saw all objects as essentially the shapes of classical geometry (for they believed with Plato that God eternally geometrizes), they recognized that the relationship between the shape of an object as it was and the shape as it appeared to the eye from varying angles of vision must be expressible in terms of mathematics. They worked out various independent and disconnected theorems of perspective and gave to the geometry that grew out of their work its two basic terms: *projection* and *section,* the latter referring to the point of view, or eye of the painter, from which an object or a group of objects is viewed; and the former, to an imaginary plane which intercepts or cuts that view, and is the picture itself.

Mathematically, we express these same relationships in the following manner:

From a point O, lines are drawn to every point of a geometric figure F; these lines issuing from O are cut by a plane w. . . . The set of lines joining a point O to the points of a figure F is called *the projection of F from O*. If a set of lines issuing from a point O is cut by a plane w, the set of points in which the plane w

118

cuts the lines through O is called *the section of the lines through O by the plane w.*

Let us observe now what this definition means in relation to the projection and section of the circle.

We begin by drawing a circle and selecting a point O in space above the circle and directly above the center. From every point on the circumference of the circle we conceive a line joining that point and O and continuing on past O. We now have in our mind a set of lines forming an infinite double cone. This is the *projection* from O of our original circle. If we conceive of a plane surface cutting this projection in various ways, we shall have a series of *sections of the projection.* A section which cuts one portion of the cone will give us a picture of our circle as an ellipse. A section which is made parallel to any line of the cone will give us a parabola; and a section which cuts both portions of the cone, a hyperbola (page 120).

Here is a beautiful example of the elegant generality of this new geometrical way of thinking. Menaechmus and Apollonius studied the conics under the almost unbearable weight of a cumbersome terminology. Descartes dealt with them as varying forms of the general equation of the second degree in two unknowns. Projective geometry now enables us to define the conic sections with even more stunning simplicity:

The conic sections are simply the projections of a circle on a plane.

The artist's yet unorganized mathematics of perspective and the mathematician's yet undiscovered art of projective geometry met in a man named Gérard Desargues (1593-1662), a self-educated architect and engineer. Desargues' interest in the subject was purely practical. "I freely confess," he wrote, "that I never had

119

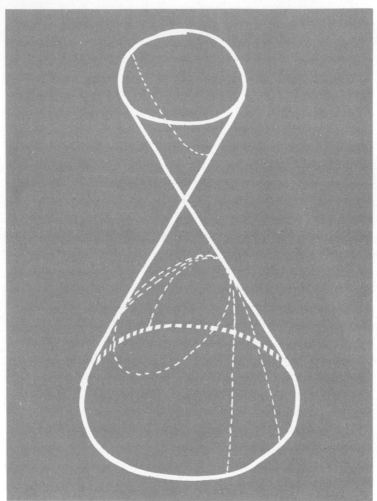

taste for study or research either in physics or geometry except in so far as they could serve as a means of arriving at some sort of knowledge of the proximate causes . . . for the good and convenience of life, in maintaining health, in the practice of some art. . . ." He began by organizing numerous useful theorems and disseminating these through lectures and handbills. Later he wrote a pamphlet on perspective which attracted very little attention. His chief contribution, the foundation of projective geometry, appeared in 1639. With a few important exceptions, it was entirely ignored by his contemporaries; and every printed copy was lost. Only by the chance discovery of a manuscript copy, two hundred years later, was Desargues' original contribution to mathematics known at all.

The difference between Desargues' geometry and Euclid's can be seen most vividly in the figures as they are presented by each man. Euclid is concerned with showing that two figures are *congruent*. This means that if we could slide a given figure by what a geometer calls "rigid motion" (which causes no change in the figure during the moving) along the plane to a second figure, the two would coincide. In the language that the new projective geometry was to bring to mathematics, we say that Euclid was concerned with those characteristics of the figure which are *invariant* (do not change) under the transformation of rigid motion:

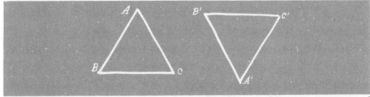

Euclid was also concerned with those characteristics of the figure which are invariant under the transformations of uniform expansion or condensation:

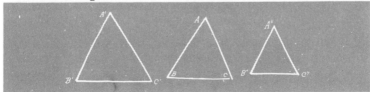

Euclid, of course, did not himself think of his concerns in these terms. Given two triangles, he took as his object to show that they were *congruent*. (They were, for instance, if all three sides of the second were equal to the corresponding sides of the first.) If they were not congruent, his object was to show that they were *similar*. (They were similar if all three corresponding angles were equal.)

The one theorem in projective geometry which bears the name of Desargues shows immediately by its wording and the accompanying figure that Desargues was concerned with relationships between triangles quite different from those that had Euclid's attention:

THEOREM: *If in a plane two triangles ABC and A'B'C' are situated so that the straight lines joining corresponding vertices meet in a point O* (in the language of art, are in perspective from O), *then the corresponding sides, if extended, will intersect in three collinear points QRP.*

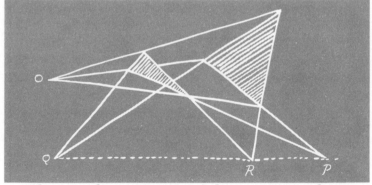

The triangles in Desargues' theorem are neither congruent (sides not equal) nor similar (angles not equal); yet there exists between them a relationship, as stated by the theorem above, which does not change—remains invariant—under the transforming powers of projection. The reader can test this statement experimentally by

122

drawing other figures to illustrate the theorem. With one exception, which we shall take up later in this chapter, he will find that the theorem always holds.

Desargues' excitingly new theorem and his really revolutionary little book were taken seriously by few people. A mapmaker named Philippe de la Hire, who had been one of his pupils for a time, utilized the new ideas of projection in his work and made a careful manuscript copy of Desargues' work, which saved it for posterity. Another fellow countryman, a youthful genius named Blaise Pascal (1623-1662), using the method of projection, proved what has been called "one of the most beautiful theorems in the whole range of geometry." (We shall state this theorem later, on page 124.) With these two exceptions, projective geometry, which was invented by Desargues in 1639, might just as well have not been invented until the beginning of the nineteenth century, when it was invented all over again!

The story of the second invention of projective geometry is one of the most dramatic in the history of mathematics. During Napoleon's retreat from Moscow, a young officer of engineers named Jean Victor Poncelet (1788-1867) was left for dead on the battlefield. He was picked up by enemy soldiers only because they thought that being an officer he might be able to give useful information. As a prisoner of war, he was forced to march for nearly five months across frozen plains to his prison on the banks of the Volga. At first he was too exhausted, cold and hungry even to think; but when spring came ("the splendid April sun"), he resolved to utilize his time by recalling all he could of his mathematical education. Later he was to apologize that "deprived of books and comforts of all sorts, distressed above all by the misfortunes of my country and my own lot, I was not able to bring these studies to a proper perfection." Nevertheless, a year and a half later, he returned to his native France, carrying with him the notebooks which were to serve as a passport for all mathematicians to "the enchanted realm." He was twenty-four years old at the time.

In his classic treatise on projective geometry, published in 1822, Poncelet introduced a convention which has been used in all textbooks on the subject since his time. This was a simple typographical arrangement which brings immediately to the eye "the enchanted realm, where thought is double."

Point and line in plane projective geometry are called dual elements. Drawing a line through a point and marking a point on a line are dual operations. Two figures are said to be dual if they can be obtained each from the other by replacing every element and operation by its dual element and operation. Two theorems are called duals if one becomes the other when all elements and operations are replaced by their duals.

Poncelet emphasized this distinguishing duality of thought in projective geometry by displaying all theorems in pairs. Thus the beautiful theorem of Blaise Pascal which we mentioned earlier in this chapter is displayed beside its dual, a theorem proved much later by C. J. Brianchon (1785-1864):*

PASCAL'S THEOREM	BRIANCHON'S THEOREM
If the *vertices* of a hexagon *lie alternately on* two straight *lines*, the *points where opposite sides meet* are *collinear*.	If the *sides* of a hexagon *pass alternately through* two *points*, the *lines joining opposite vertices* are *concurrent*.

* The hexagons referred to in these theorems are figures formed when any six points are joined serially. The reader may enjoy joining six straws of varying lengths and discovering the varied hexagons he will obtain that way.

124

The duality of the two theorems stated above becomes even more vivid when we list parallel terms in parallel columns:

PASCAL'S THEOREM	BRIANCHON'S THEOREM
vertices	sides
lie alternately on lines	pass alternately through points
points	lines
where opposite sides meet	joining opposite vertices
collinear	concurrent

These theorems do not show an obvious resemblance; yet they are as firmly linked as Siamese twins.

Pascal's theorem was proved in about 1639, before his sixteenth birthday; Brianchon's was discovered—through the principle of duality—while he was a student at the École Polytechnique, and was printed in the school *Journal* in 1806—when Brianchon was twenty-one.

According to the Principle of Duality, *the dual of any true theorem of projective geometry is likewise a true theorem of projective geometry.*

Projective geometry is indeed an enchanted realm—a sort of Big Rock Candy Mountain of mathematics—where every theorem yields a twin and the proof of the first provides, with the proper exchange of dual elements and operations, the proof of its twin. "Thought is double and flows in parallel streams."

There is, however, a truly marvelous paradox in this world of parallel thought. For the beautiful and completely general duality of projective geometry depends upon the fact that in projective geometry *there are no parallel lines.*

In the elimination of parallel lines, projective geometry makes a complete break with its parent art, perspective. When we are drawing a scene, we draw those lines

125

which are parallel to the frame (the pillars at either side, the edge of the floor or the table top) parallel. This is in spite of the fact that we never actually see these lines as truly parallel, for all parallel lines appear to the eye to be approaching each other. In the small area framed by the picture this optical illusion, however, is not usually apparent.

The simple sketch below shows two different treatments of the parallel lines in the scene. The pillars at the sides are parallel to each other and to the sides of the frame. The horizontal lines of the tile floor are parallel to each other and to the upper and lower sides of the frame. All other "parallel" lines in the picture appear to be approaching a single "vanishing point" on the horizon.

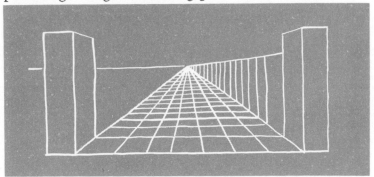

In his great work on projective geometry, however, Poncelet proposed a convention which would do away entirely with parallel lines in the mathematics that was the child of perspective. What he proposed was very simply to expel parallel lines from projective geometry *by fiat*. Although his principle as stated sounds as metaphysical as anything in mathematics, his purpose was merely the practical, down-to-earth one of eliminating bothersome exceptions always having to be made in theorems and proofs for the special case of parallel lines.

126

Let us recall as an example Desargues' theorem, which we stated earlier on page 122. We illustrated this theorem with a figure similar to the one below, and we observed that it is indeed true that the extended corresponding sides of the two triangles meet in pairs in three collinear points. But now let us consider a slightly modified figure where one side of one triangle is parallel to the corresponding side of the other triangle:

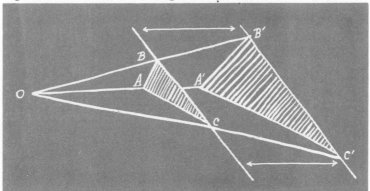

Although we can never hope to examine these lines in their entirety, it is immediately clear to us that the lines *BC* and *B'C'* will never meet. We cannot make this statement on the evidence of our eyes, for—if we extend the lines far enough—our eyes will tell us that they are indeed approaching each other and must, therefore, eventually meet. We make this statement because we *know* that parallel lines will never meet, "because," we say, "that is what parallel lines are—lines that never meet."

But what shall we do about the statement of Desargues' theorem that the corresponding sides of the triangles, if extended, will meet in three collinear points? We shall have to add a qualifying clause to the theorem, "unless the corresponding sides are parallel."

It would be one thing if Desargues' theorem were the

only one to which we had to add such a clause to cover the exceptional case of parallel lines; but it is not. Almost every principle, theorem and proof in projective geometry must be modified to cover the exceptional case of the parallel. Such modifications are repugnant to mathematicians. Economy is one of the prime requirements for beautiful, general and effective mathematics. A theorem which applies to one specific triangle only is no theorem at all. A theorem that applies to almost all triangles is an improvement. But a theorem that applies to all possible triangles without exception—now there is a theorem to delight a mathematician!

Obviously, life in the enchanted realm of projective geometry would be much better for mathematicians—and for mathematics—if there were no parallel lines! This is exactly what Poncelet proposed to accomplish.

It is an axiom of ordinary geometry that any two straight lines (except two parallel lines) intersect at one, and only one, point. If we now postulate that any two parallel lines have one *ideal* point in common, then we can state this important axiom with even greater generality:

Any two lines meet in one, and only one, point.

In the case of non-parallel lines the intersection is a real point; in the case of parallel lines it is an ideal point. But this distinction is trivial compared to the fact that the axiom now applies to *all* lines without exception.

Unfortunately, as we have seen in the extension of the concept of number, the use of such words as "real" and "imaginary" and now "ideal" is often, even with mathematicians, a great hindrance to the grasping of a new idea. If, as is often done today, we simply postulated in our mathematics the existence of two kinds of lines

128

and two kinds of points, we might escape this language trap. Any pair of lines, we might say, meets in one and only one point. Whether this point is of Class A or Class B depends upon the class to which the pair of lines belongs. Removed from the crippling language of everyday life, we might pursue our object with logic alone.

The language of everyday life, however, is not completely crippling even to mathematics. In fact, to switch metaphors, it often provides us with a very useful crutch in developing new ideas. Because parallel lines appear to meet or to be approaching a meeting place at the horizon, we say in our mathematics that parallel lines meet "at infinity." Since all parallel lines with a common direction are conceived as having an ideal point in common, we also conceive of all these ideal points of all possible sets of parallel lines as being on an ideal line, "the line at infinity." The mathematician uses this language very much as a poet uses a metaphor. Although he can make this principle premise in an analytic basis, he finds the language of everyday life both simple and suggestive in handling these new ideas.

Most of us have forgotten that we are doing very much the same thing when we talk about the "real" points of ordinary geometry. A "real" mathematical point is the *idealization* of a real everyday point made with a pencil or a pen. This point, no matter how carefully we make it, has of necessity dimension. In fact, it has three dimensions—length, breadth and a certain theoretically measurable depth when it is made with a pencil or a pen. Our so-called "real" point of geometry has no dimension at all. Yet we easily conceive of a real line in our geometry, the length of which is composed of an infinite number of these dimensionless points!

In the familiar mathematics of everyday life we are

working with ideas that are every bit as "far out" as the concept of parallel lines meeting in ideal points that lie on an ideal line—at infinity. Although we are probably not aware of the fact, we have been forced to accept these ideas as logical necessities. Without the axiomatic statement that a line is composed of an infinite number of points and that there is a unique number for every point on a line, neither analytic geometry nor the calculus, to mention only two examples, would work as effectively as they do—in practical problems as well as in higher mathematics.

It works. It is a logical extension of and logically consistent with our basic principles. *And it works.* This is justification enough for a mathematician to incorporate into projective geometry a postulate which eliminates the nagging exceptional case of the parallel. Just as we can postulate that any two non-parallel lines meet in one, and only one, point (in spite of the fact that the actual point in which they intersect covers an infinite number of mathematical points), in the same way we can postulate that any pair of parallel lines have in common one ideal point (in spite of the fact that they only appear to our eye to meet).

We have already seen that when we accept this principle, the statement of Desargues' theorem no longer requires an exception for the special case where any pair of sides of the two triangles is parallel. But the simplest example of the way in which this principle allows mathematicians to unify and generalize projective geometry lies in the concept of projection itself. Originally it was necessary to distinguish between different types of projection, one in which the lines from the points of the figure meet in a single point and one in which the lines are parallel to one another:

130

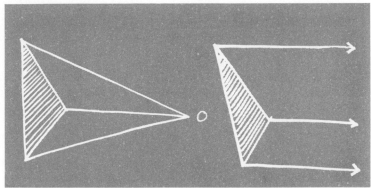

When we examine these projections in the figure above, they appear to be completely different. But if we conceive of O, the point of projection in the left-hand diagram, as moving away from the figure toward infinity, it is clear that as O approaches infinity the lines joining the points of the figure to O will become more and more "parallel."

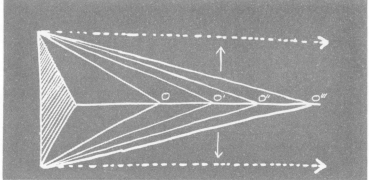

Utilizing our new concept of ideal points, we can say that both projections are from a point O. In the left-hand diagram above, O is a real point; and in the right-hand diagram, it is an ideal point. We can now discuss both projections without distinguishing between them as special cases, because they are both projections from a single point!

131

Such elegant economy is a prized virtue in mathematics. It makes for practicality as well as for beauty.

A realm where thought is double and flows in parallel streams—where the statement of every theorem and its proof automatically yields another true theorem and another proof—this is enchanting economy. A realm, however, where exceptions must be made in every theorem and every proof is under an anti-mathematical spell. Mathematicians have been exorcising such spells for two thousand years, and will continue to do so!

To TRISECT A GIVEN ANGLE.

Mathematicians, amateur and professional alike, have struggled with this simple-sounding problem. Plato as well as Archimedes tried to trisect the angle. Another man, two thousand years later, wrote in his autobiography: "When I reached geometry, and became acquainted with the proposition the proof of which has been sought for centuries, I felt irresistibly impelled to try my powers at its discovery." Any mathematically inclined person will recognize this response. We have all *tried* to trisect the angle.

The trisection of the angle was one of the four great construction problems that the Greeks left to mathematics, the other three being the doubling of the cube, the squaring of the circle, and the construction of a polygon—other than triangle and pentagon—with a prime number of equal sides. From a practical point of view these constructions are not too difficult. With a protractor and a ruler we can draw what will appear to be a quite perfect regular heptagon. We can make a square having essentially the same area as a given circle and a cube having essentially twice the volume of a given cube. With protractor and ruler we can also divide any given angle into three "equal" parts—parts

*The
Possibility
of Impossibility*

133

which, for all practical purposes, will be quite equal.

The protractor and ruler we use for these constructions have, however, one thing in common which would make them repugnant to Greek eyes. They are both measuring devices. The protractor measures off the circular angle in degrees, minutes and seconds; and the ruler measures off lengths in units and parts of units. They are both very useful instruments, but there *is* a certain mean practicality about them. To the eye of man, constructions made with such instruments might appear accurate, but the gods would know different. The Greeks—and the gods —were not interested in the practical construction of squared circles, doubled cubes or trisected angles. They were interested in constructions which would in theory be absolutely exact even though in practice, because of limitations inherent in man and his instruments, they would be indistinguishable from the approximated constructions by ruler and protractor.

Although no mechanical device can possibly mark off on a line the exact point which is the irrational distance from the beginning represented by $\sqrt{2}$, we can in theory mark off *the exact distance* by constructing a right triangle of unit size and swinging across the number line an arc, the radius of which is the length of the hypotenuse, or $\sqrt{2}$. This arc actually marks $\sqrt{2}$ no more exactly from a practical point of view than an ordinarily good ruler would; but—in principle—it is exact. If the number line could be represented by an infinite number of points and if the compass could trace the path of just one point at all positions the same distance ($\sqrt{2}$) from the point 0 on the line, this path would of necessity intersect with our right triangle at the point which is the vertex and the number line, or extended base of the right triangle, at the point which is the distance $\sqrt{2}$ from the origin.

134

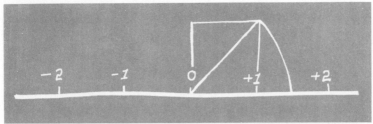

The reader will note that for this "theoretical" construction of √2, we have used no measuring device like protractor or ruler. We have assumed, with Euclid, that from a given point 0 we can draw a straight line to another point 1. (We do not have to measure this distance, since any distance we choose can serve as our unit.) We have also assumed, with Euclid again, that we can extend a given straight line and that we can draw a circle with given center and radius. The construction of the isosceles right triangle on a given base is Proposition 10 of Book I of the *Elements*.

For our construction then we have used only an unmarked straightedge and a compass. These, being the mechanical manifestations of the straight line and the circle, were, as far as classical Greek mathematics was concerned, the only instruments which could be used in construction. The traditional problems thus were:

To construct by straightedge and compass alone:
> A regular heptagon.
> A square equal in area to a given circle.
> A cube double the size of a given cube.
> An angle one-third of a given angle.

It was the restriction to straightedge and compass alone which made these problems "problems."

Even if we eliminate the crass idea of marked-off measure but allow an instrument other than straightedge

135

and compass, we can make all of these constructions. As an example, we have already seen in Chapter 5 that the problem of doubling the cube, or the solution of the equation

$$x = \sqrt[3]{2}$$

can be determined by the intersection of conic sections which require only simple mechanical instruments for their construction. Without the restriction to straightedge and compass, there would have been *no* classic construction problems.

It is impossible at this date even to estimate the mathematical man-hours that have been devoted to the classic construction problems. For more than two thousand years every mathematician born in the Western world has had his turn at one or all of them. New mechanical devices have been invented, new curves have been discovered, new branches of mathematics have been developed, all in the course of efforts to solve these problems. Yet on the eve of the eighteenth century all four of them still stood, absolutely undented. Their hour, however, had at last arrived.

In the long assault there had always been an unstated and equally unquestioned assumption on the part of the mathematicians who tackled the problems. Everybody assumed that *it was possible* to construct a regular heptagon, to square a circle, to double a cube, and to trisect an angle—with straightedge and compass alone. In 1796 a young man, just nineteen, became the first person in the history of mathematics to question this age-old assumption. Karl Friedrich Gauss considered an entirely new idea: *perhaps it is impossible to construct these figures under the classic restriction.*

The possibility of impossibility!

It was a revolutionary idea. Up to the beginning of

136

the nineteenth century, in the history of mathematics there had been only one other comparable thought. That was when the Pythagorean, pondering the diagonal of the unit square, considered the possibility that there might be no rational number which when multiplied by itself would produce 2.

The young Gauss was particularly interested in just one of the classic problems, the construction of the regular polygons. The Greeks, some two thousand years before him, had constructed within the circle the equilateral triangle, the square and the regular pentagon. From these basic figures they had gone on to construct the regular hexagon, octagon, decagon and 15-gon, the number of sides of which in each case is a product of the basic 2, 3 and 5 of triangle, square and pentagon. It was clear that, by continuing to bisect the sides of these polygons, they could produce a 12-gon, 16-gon, 20-gon, 30-gon and so on. *But could they produce a regular heptagon* (7 sides) *with straightedge and compass alone?* This the Greeks left as an exercise for the future; and the future—up until the time that the young Gauss entered it—had produced neither a regular heptagon nor a single regular polygon the construction of which had not already been known to the Greeks.

Gauss, however, began with great advantages over the ancients. He had a language, algebra, and a tool, analytic geometry, which allowed him to attack the problem in a much more general way than had been possible for them. Although all of the construction problems are presented differently—some even, like the Delian problem, with a story to go with them—they are, in the language of algebra, essentially the same: certain lengths are considered to be given, and one or more lengths must be found. To solve a given problem, we must find a relation between the unknown quantities $(x, y, z, \ldots)$ and the known

137

quantities $(a, b, c, \ldots)$. We must state this relation as an equation; and then—and here is the crux of the matter —we must determine whether the solution to this equation can be obtained by algebraic processes which are the equivalent of straightedge and compass constructions.

At first we may be set back by the idea of algebraic processes as geometric constructions with straightedge and compass; but a moment's thought will assure us that we have thought for a long time in this manner. It is clear that, taking two segments of lengths a and b (in terms of a given unit segment), the solutions to such simplified equations as

$$a + b = x \qquad \text{or} \qquad a - b = x$$

can be found with these traditional instruments:

$$\longleftarrow a \longrightarrow \quad \leftarrow b \rightarrow \qquad\qquad \longleftarrow\!\!\longrightarrow a \longrightarrow$$

$$\qquad\qquad\qquad\qquad\qquad\qquad\qquad\qquad \leftarrow b \rightarrow$$

$$\longleftarrow x \longrightarrow \qquad\qquad\qquad \longleftarrow x \longrightarrow$$

$$a + b \qquad\qquad\qquad\qquad a - b$$

It is not quite so immediately clear that we can also solve such equations as

$$ab = x \qquad \text{or} \qquad a \div b = x$$

with similar constructions. Yet these too are possible:

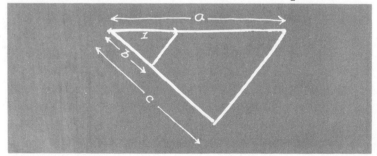

Utilizing the fact that in both of these examples we have constructed similar triangles with one side of the smaller triangle as the unit, we can show that in the multiplication problem illustrated above:

$$\frac{1}{b} = \frac{a}{c} = \frac{a}{ab}$$

the segment c being the desired segment ab. On the other hand, in the division problem, which is illustrated in the figure below, we can determine

$$\frac{a}{b} = \frac{c}{1} = \frac{\frac{a}{b}}{1}$$

the segment c being the desired segment a/b, or $a \div b$:

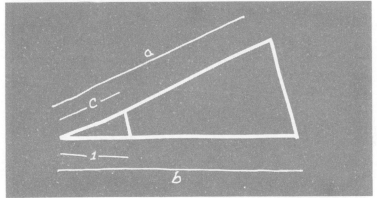

From these simplified examples it is clear that the rational operations of algebra—addition, subtraction, multiplication and division—can all be performed by geometrical constructions which require only straightedge and compass. It follows that any equation which *can be solved* by any finite combination of one or more of these processes *can also be constructed* by straightedge and compass alone. (It must be a finite combination because obviously if the number of operations required were infinite we would never be able to finish the construction.)

Besides the four basic operations of addition, subtraction, multiplication and division, there is one other operation in algebra which is the equivalent of a construction by straightedge and compass alone. That is the extraction of square root. Given the equation

$$x = \sqrt{a}$$

we can solve for x in the following manner:

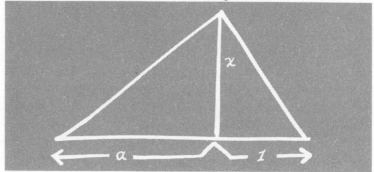

After establishing the similarity of the triangles in this figure, we can conclude

$$\frac{a}{x} = \frac{x}{1}$$
$$x^2 = a$$
$$x = \sqrt{a}$$

It can be shown that the solutions for x which can be obtained by any finite number of additions, subtractions, multiplications, divisions and extractions of square root include *all possible segments from a given set which can be constructed by straightedge and compass alone.* There is nothing at all mysterious about this relationship between the solution of equations and the construction of geometric figures. We need recall only the fact that a straight line and a circle are represented in analytic geometry by equations of the first and second degree, respectively, and that the determination of circles with straight lines, or with other circles, leads analytically to the solution of equations which involve no irrational operations other than the extraction of square roots.

Herein lies a method for establishing that a given construction problem is impossible if the tools of construction are restricted in the classic manner. All we have to do is to show that the problem requires the construction of a segment which cannot be obtained from the measure of the given segments by straightedge and compass; i.e., the solution of an equation which cannot be obtained by the four basic operations and the extraction of square root. This is, naturally, not so easy as it sounds. Yet, one by one, the famous construction problems of antiquity, which withstood so firmly the full arsenal of two millenniums of mathematics, have fallen before this new approach, called the algebra of number fields.

The first problem to be toppled—by the young Gauss himself—was that of constructing a regular heptagon with straightedge and compass alone. Such a construction, Gauss showed, is impossible because, unlike the pentagon, it results in a cubic equation the solution of which cannot be obtained by the four rational processes and the extraction of square root. In the course of showing that the required construction of a regular heptagon is impossible, he established the fact that the only constructible regular polygons with a prime number of sides are those with p sides where p is a prime of the form $2^{2n} + 1$. The first such constructible regular polygon after the triangle and the pentagon of the Greeks is the 17-gon $(2^{2^2} + 1)$. Gauss's general proof, which established the conditions

141

for constructibility of the regular polygons and provided a tool for attacking the other construction problems, was a magnificent achievement. Even Gauss himself was impressed by it. He had been torn between a career in philology and one in mathematics, but now he definitely decided in favor of mathematics. When the score is added for the classic construction problems—time spent against the advantages accrued—the recruitment of Gauss must weigh heavily.

Last of the problems to topple was the famous question of squaring the circle. Almost a century after Gauss's solution of the problem of the regular polygons, Ferdinand Lindemann (1852-1939) succeeded in proving that π cannot be the solution of an algebraic equation with rational coefficients. Since all constructions by straightedge and compass *can* be represented by equations with rational coefficients, this indirectly established the impossibility of squaring the circle, or solving the equation

$$x^2 = \pi r^2$$

In the century between Gauss and Lindemann, the other two problems yielded almost automatically. Both are impossible under the classic restriction. We have already seen that the solution of the Delian equation, $x^3 = 2$, involves the extraction of a cube root; and we shall now examine the proof that in general the trisection of the angle is also impossible by means of straightedge and compass alone and for the same reason.

We begin by inscribing on the complex plane a unit circle with center at O and an arbitrary angle with vertex at O and one side lying along the real axis. The point where the arbitrary side of the angle cuts the unit circle is represented by the complex number Θ:

142

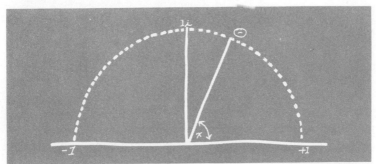

This complex number, as we recall from Chapter 7, is of the general form $x + iy$, where x and y are real numbers and $i = \sqrt{-1}$. It is uniquely determined by its distance from the point of origin and by its angle with the positive side of the real axis. These two characteristics are called, respectively, the *absolute value* and the *argument* of the complex number $\ominus$:

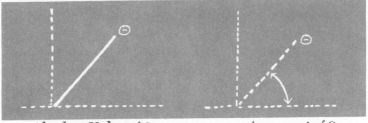

Absolute Value of $\ominus$ *Argument of* $\ominus$

    In the geometric interpretation of complex numbers, multiplication of two complex numbers is defined as the product of the absolute values and the sum of the arguments. Since the absolute value of $\ominus$ is 1, any root of $\ominus$ will be a complex number on the circumference of the unit circle, all of which also have an absolute value of 1.* Its exact location on the circumference must be determined

* This is easily established by the theorem of Pythagoras.

by the argument, or size of the angle it makes with the real axis. The square root of Θ, for example, will be that point, or complex number, where the *bisected* angle x cuts the circumference. This is an operation which we can perform with straightedge and compass alone:

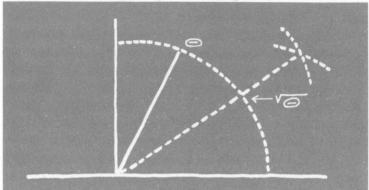

The cube root of Θ will also be a complex number on the circumference, one that makes an angle with the real axis equal to one-third of angle x, or a *trisected* angle x. We cannot, however, locate this number as we located the square root of Θ because it is impossible to extract the cube root of a complex number by algebraic operations which correspond to construction by straightedge and compass alone. It is, therefore, impossible to trisect a given angle under the classic restriction which the Greeks placed upon the problem, just as it is impossible to construct a regular heptagon, square a circle or double a cube.

That should settle the question for all time, but there is a psychological epilogue to the proofs that each of these famous problems is impossible. Mathematicians, amateur and professional alike, have shown a great reluctance to part with their old friends. Even the great Irish mathematician William Rowan Hamilton (1805-1865) wrote to De Morgan as late as 1852: "Are you sure that it is impossible to trisect the angle by Euclid [*i.e., under the restriction*]? I fancy that it is rather a tact, a feeling, than a proof, which makes us think that the thing cannot be done. But would Gauss's inscription of the regular polygon of seventeen sides have seemed, a century ago, much less an impossible thing, by line and circle?"

This is curiously emotional language from a mathematician, especially when the essence of Gauss's proof is not the possibility of constructing a regular 17-gon but the impossibility of constructing a regular heptagon. Apparently the impossible is hard for any of us to accept. It seems almost a personal challenge, and this feeling is perhaps responsible for the fact that in spite of the finality of mathematical proof that the things cannot be done, would-be constructors of regular heptagons, squared circles, doubled cubes and trisected angles continue with us well into the twentieth century. Any statement in print that one of the problems is impossible invariably brings to the author a beautifully drawn construction, usually with protractor and ruler, with a modest request for "comment."

Why have these famous problems captured the general imagination so permanently? Perhaps because, stated as they are in the language of construction, they have a practical sound which is refreshingly removed from the abstractions of most higher mathematics. This is ironic— for in these problems no one, including the Athenians who consulted the oracle, was ever concerned with the actual construction of anything. Even Gauss's famous proof that it is possible to construct by straightedge and compass alone a 17-sided regular polygon did not show how to construct such a polygon.* The truth of the matter is that

* A simple method of constructing the regular 17-gon is given by H. S. M. Coxeter in his *Introduction to Geometry* (New York: John Wiley and Sons, Inc., 1961).

the construction problems, in spite of their practical sound, are as highly artificial as any mathematical problems can be.

It is indeed a curious thing that mathematics would hobble itself with an impossible restriction and then spend two thousand years trying to construct regular heptagons, squared circles, doubled cubes and trisected angles which could be constructed in a trice with a reasonably accurate protractor and ruler. But it was fun and—mathematically speaking—it was extremely profitable fun. Asked, "Was it worth it?" mathematics as a whole, immeasurably enriched by the discovery of the conic sections, the invention of analytic geometry, the winning over of Karl Friedrich Gauss to mathematics, the algebra of number fields, would echo with Hamilton: "I have not to lament a single hour thrown away on this attempt."

FOR THE READER

It was Augustus De Morgan, the great mathematical writer of the last century, who mourned the Greek limitation to straight line and circle:

"What distinguishes the straight line and circle more than anything else, and properly separates them for the purpose of elementary geometry? Their self-similarity. Every inch of a straight line coincides with every other inch, and of a circle with every other of the same circle. Where, then, did Euclid fail? In not introducing the third curve which has the same property—the *screw*. The right line, the circle, the screw—the representation of translation, rotation, and the two combined—ought to have been the instruments of geometry. With a screw we should never have heard of the impossibility of trisecting an angle, squaring a circle, etc."

Let us take a moment to examine how De Morgan's proposed inclusion of the mathematical screw, or helix, as an instrument of construction would allow us to trisect an angle. Since the helix makes one complete turn in its length, the angle of the screw thread is proportional to the length of the shank; one-third of a complete turn of the screw would require one-third of the length of shank necessary for a full turn. The problem of trisecting any given angle would then be merely one of obtaining a segment one-third of a given length of the shank. This we could easily do; for while we cannot trisect an angle with straightedge and compass alone, we can trisect a line.

To divide a given segment into three parts, we construct an angle with the given segment as one side. We mark off the unit three times in succession on the other side. We join the point which marks the end of the third unit with the end of the given segment and join the ends of the other unit lengths to the given segment by parallel lines. In this way we have constructed three similar triangles, the corresponding sides of which are in the same ratio. Since the segment AC is divided into unit thirds, the given segment AB must also be divided into thirds.

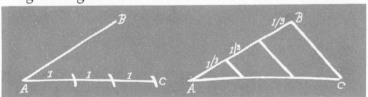

Using this same method, the reader should try dividing an arbitrary segment into sevenths.

10

*Euclid
Not Alone*

THE SUBJECT OF GEOMETRY IS ALMOST synonymous with the name of Euclid. For this reason, when we first hear of something called non-Euclidean geometry, we feel that there is some misunderstanding. Why, Euclid *is* geometry! But our trouble is only in our tenses. Euclid *was* geometry for more than two thousand years. He isn't any more.

The story of how Euclid was deposed, and at the same time elevated, is one of the longest, in many ways the most ironic, and without question one of the most important in the history of mathematics.

As we recall from Chapter 2, Euclid deduced all of his theorems, or propositions as they were sometimes labeled, from a relatively small set of definitions and basic assumptions, called, more or less interchangeably, *axioms* or *postulates*. For a very long time it was believed that these assumptions of Euclid's, which we have printed in full on page 27, were true, in the ordinary way of what we mean by "true"; and because they were true, the theorems which were logically deduced from them were "true" in the same ordinary way.

Yet geometry is a subject whose "truth" is immediately controverted by its very name. *Geometry* means *earth-measurement*, and that was an accu-

148

rate name for the art which the Greeks learned from the Egyptians. On the small part of the earth which was flooded each year by the Nile, the Egyptians found it necessary to develop a system of measurement by which they could reestablish boundary lines after each inundation. But let us take a globe—for the earth itself, as we shall see, is for various reasons too large for our purposes —and let us take a few of the "truths" which the Egyptians arrived at from experience and which the Greeks deduced in logical fashion from their axioms and postulates.

A straight line is the shortest distance between two points.

The sum of the angles of a triangle is 180°.

The circumference of a circle is $2\pi r$.

These ideas of straight lines, triangles and circles are almost as familiar as our own faces. We all know, for instance, what a straight line is. It is the shortest distance between two points, and it is, well, straight. But when we try to draw a straight line on the surface of the globe, it is immediately apparent that we can't draw any sort of line which even begins to meet our intuitive idea of what a straight line should be. Obviously (it is not at all obvious, but we think it is!), we can stretch a thread across the surface of the globe between any two points (say, San Francisco and London), and find the shortest distance between them. Since "the shortest distance between two points" satisfies part of our definition, we can call the line marked by the thread a straight line if we will just forget what we usually mean by straight. If we extend the line which marks the shortest distance between San Francisco and London all the way around the globe, we find that it divides the surface into two equal parts. In other words, it is a great circle. The great circle with which we are most familiar is the one we call the equator. Although arcs of these great circles are the straight line segments of our surface—being the shortest distance between two points—our idea of straightness is violated by calling them such, and so we call them the *geodesics* of the surface. The geodesics of the Euclidean plane, or a perfectly flat surface like a floor, are what we call "straight" lines.

Since we cannot draw "straight" lines on our globe, we cannot have straight-sided triangles. Our triangles will bulge on the sides and in the center. If we take one such triangle, flatten it with as little distortion as possible onto this page, and then join its vertices with straight lines, we see at a glance that if the sum of the angles of the interior triangle is 180°, as we know by Euclidean geometry that it is, the sum of the angles of the spherical triangle must be more than 180°.

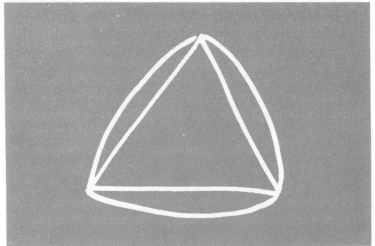

We have seen that the shortest distance between two points on the globe is not a straight line, that the sum of the angles of a triangle on the globe is not 180°. Now let us draw a circle on our globe. It meets exactly the Euclidean definition of a circle as "the locus of points equidistant from a center," and we may jump to the conclusion that all we know about a Euclidean circle will also be true of such a circle. But the interior of this circle does not look like the interior of the circle we know about. It is two-dimensional, but it is not flat. It may look, depending on how large it is and how large is the globe upon which it is drawn, like a ball cut in half or a hub cap or merely a saucer. If we place it on this page and trace around its edge, we shall have a flat circle. Of this circle we know that the circumference is twice the product of π and the radius. But obviously the curved circle drawn on the surface of the globe, which must have had the same circumference, cannot have had the same radius. Its radius must have been greater because of the curvature of the surface on which it was drawn. Its circumference, therefore, cannot be equal to $2\pi r$.

Although geometry means earth-measurement, it is apparent that the measurement of the earth has very little to do with the geometry of the Euclidean plane. This was not because the Greeks of Euclid's time (300 years before Christ) did not know that the earth was round. They had calculated that it was, from the fact that the North Star was higher in Greece than it was in Egypt. But the geometrical figures on which they based their geometry were drawn on only a small part of the surface of the earth, and that part, for all practical purposes, was flat. It would be more exact to say that they based their geometry on idealized figures on an ideal plane, and these were only represented by those which they drew on the earth. Euclid's geometry was indeed, as Edna St. Vincent Millay has written, "nothing, intricately drawn nowhere."

Yet for two thousand years, in spite of the fact that

the geometry of Euclid did not truly apply to the only large surface which man knew and had not constructed himself, it was felt that this geometry—then the *only* geometry—represented "truth," in so far as man could know it. One philosopher (Kant) called the ideas from which Euclid deduced his theorems "the immutable truths"; another (Mill) considered them "experimental facts." Mapmakers and sailors might struggle with the geometry of the boundless, finite surface that is our planet; but Euclid's geometry, extended to three dimensions and a space which was thought both boundless and infinite, was the geometry of God's mind.

That the geometry of Euclid was not the only one possible, either physically or mathematically; that it was deduced not from self-evident truths but from arbitrarily chosen and unprovable assumptions; that another choice of assumptions could yield a geometry just as consistent, just as useful and just as true, never occurred to anyone for more than two thousand years unless, in a sense, it had occurred to Euclid himself when he set out the assumptions on which he based his geometry. For today it is clear that Euclid recognized what no other man between his time and that of Gauss recognized: *that his axioms and postulates were assumptions which could not be proved.*

The idea of those who followed Euclid and extolled him was that the axioms and postulates of his geometry did not have to be proved because they were self-evident. There was only one impediment to the full and complete acceptance of this point of view and that was the fifth postulate, which makes a statement very roughly equivalent to our common statement that parallel lines never meet. From the beginning, compared to the other axioms and postulates, this one did not seem quite self-evident enough, even to the most devoted admirers of the master.

152

The famous fifth postulate stated as follows:

If a straight line falling on two straight lines makes the interior angles on the same side less than two right angles, the two straight lines, if produced indefinitely, will meet on that side on which the angles are less than two right angles.

As J. L. Coolidge has remarked in his *History of Geometrical Methods*, ". . . . whatever else this postulate may be, self-evident it is not."

The fifth postulate makes a statement about the entire length of a straight line, a statement which can never by its nature be verified by experiment. To remove this flaw from the work of the master, generation after generation of mathematicians attempted to prove the statement about parallel lines from the other postulates. Time after time they failed: they were never able to prove the fifth postulate without substituting for it still another postulate, which simply varied the problem rather than solving it.

Among the last of the attempts to "free Euclid from every flaw" was one made by a Jesuit priest, Geronimo Saccheri (1667-1733). It was Saccheri's idea that although the parallel postulate did not, on the surface, seem as self-evident as the others, he could show that it was the only possible assumption because any other "led to absurdity." This, as we have seen, is an ancient and honorable method of mathematical proof. We assume the falsity of that which we wish to prove true, or the truth of that which we wish to prove false, and then show that such an assumption is unfeasible because it leads us to a contradiction; hence follows the truth of whatever we were trying to prove in the first place. Saccheri's method was mathematically sound; the only thing which was not sound was his attitude. When he found that assumptions about parallel lines quite different from the famous "fifth" did not

153

lead him into the expected contradictions but into a strange and fantastic geometry which was nevertheless as consistent as Euclid's, he fell back upon his feelings instead of his brains and peppered the last pages of his work with such epithets of the logically defeated as "destroys itself," "absolutely false," "repugnant." Great discoverers have made great mistakes. Columbus found the new world and at first thought that it was the old. Saccheri found a new world and refused to disembark because he thought he *knew* that there could be only one world.

It was a century after Saccheri that three mathematicians in three different countries,* independently and apparently without knowledge of Saccheri's curious contact with non-Euclidean space, came to the conclusion that Euclid had known exactly what he was doing when he made his statement about parallels a postulate instead of a theorem. He had recognized what no else had recognized: that it was completely independent of the other postulates and therefore could not possibly be deduced from them.

To prove this suspected independence of the "fifth," it was necessary only to substitute for Euclid's assumption about parallels a contrary assumption and then to show that the geometry deduced from it, in conjunction with the other postulates and axioms of Euclid, was as consistent as Euclidean geometry itself.

The first of the non-Euclidean geometries was, in the relation its axioms bore to those of Euclid, the simplest possible. All the axioms were exactly the same except one, the famous "flaw," the long-worked-over statement about parallels. We have noted that this parallel postulate may be stated in various ways, all equivalent in the sense that the same set of theorems can be deduced from any of the

* Nikolai Ivanovich Lobachevski, Russia; János Bolyai, Hungary; and Karl Friedrich Gauss, Germany.

various versions. The statement which appears in the set of axioms on page 27 is the earliest known; but since even the oldest manuscripts we have of the *Elements* date from a time nearly a thousand years after the death of Euclid, the master himself may have stated the parallel postulate in a somewhat different form. It is clear from the theorems, however, that some statement of like nature must have existed among the original set of axioms. The most easily grasped statement is a later one, known as the Postulate of the Unique Parallel:

Through any point not on a given line, one and only one *line can be drawn which will never meet the given line.*

Now let us make a contrary assumption and let us change the postulate to read:

Through any point not on a given line, infinitely many *lines can be drawn which will never meet the given line.*

Before our intuition objects to the postulate in this new form, let us recall that on the globe, where the equivalent of a straight line is a great circle, it is impossible to draw through a given point *even one line* which will never meet a given line, since every great circle intersects every other great circle. A word of caution, though. We mention the contrary example that on a sphere every straight line—or geodesic of the surface—intersects every other straight line, only to put intuition in its proper place. Mathematically, it has nothing whatsoever to do with whether the alternate above is a proper postulate.

When a set of axioms more or less agrees with our idea of reality, we will deduce from that set of axioms a geometry which also agrees pretty well with the same idea of reality. This does not mean that our idea of reality is right, but only that our axioms agree well enough with whatever reality there is so that the geometry deduced from them works.

We have seen that the earth is not the infinite plane of Euclidean geometry; yet small parts of it are, for all practical purposes, very much like small parts of the plane; and so for building pyramids and supermarkets it works very well indeed. But we shall also see that the non-Euclidean geometries, which attempted to show only from an intellectual point of view that it was possible to deduce geometries as consistent as Euclid's from a different set of assumptions, turned out to have quite a bit to do with reality, too.

The first non-Euclidean geometry, based on the same set of assumptions as the old (except for the new Postulate of Infinitely Many Parallels for the old Postulate of the Unique Parallel), applies to a surface which is the direct opposite of the surface of any part of the sphere. The surface of the sphere is what we intuitively think of as "evenly curved"; in mathematics this is more precisely defined as "constant positive curvature." The surface to which our first invented non-Euclidean geometry applies is one of "constant negative curvature." It is not (probably fortunately) a very common one in the physical world; but we can find examples of such a surface: a saddle, for instance, or a mountain pass or the surface around the hole of a doughnut. In these, however, the negative curvature is only local. For a surface of constant negative curvature, we can look ahead to the illustration on page 158.

If we place a plane tangent to a single point on a surface of constant negative curvature, like a portion of a saddle, we find that it cuts the rest of the surface in two hyperbolas. For this reason the earliest non-Euclidean geometry, which applies to such a surface of negative curvature, is called *hyperbolic* geometry. If we place a

156

plane tangent to a single point on a surface of constant positive curvature, like a portion of the sphere, and then shift the plane ever so slightly so that it is parallel to its original tangent position, we find that it cuts the surface in the shape of an ellipse. (In the special case of the sphere, it will cut a circle, which is the limiting form of an ellipse.) For this reason a later non-Euclidean geometry, which applies to such a surface of positive curvature, is called *elliptic* geometry. It substitutes for the Postulate of the Unique Parallel the following statement:

Through any point not on a given line, no line *can be drawn which will not intersect the given line.*

From our earlier experiments with our globe, we recognize that on the surface of a sphere, where a straight line is a great circle, the above postulate holds. For our purposes in this chapter, a sphere can serve as an example of a surface of elliptic geometry. Actually it is what is called "locally elliptic." To make the entire surface elliptic, a curious change must be made. As we recall, the purpose of non-Euclidean geometries is to establish the fact that geometries as consistent as Euclid's can be deduced with a different parallel postulate, the others remaining the same. It is an axiom of Euclidean geometry that two straight lines can intersect at only one point, but on the sphere two great circles always intersect at two points. To get around this difficulty, in elliptic geometry we *identify* the two points of intersection as one point. Although in this respect the geometry of the surface of the sphere as a whole is not technically elliptic and non-Euclidean, it is locally; and we can take a sphere as our sample elliptic surface.

The true surface of hyperbolic geometry—not just a portion but an entire surface—is what is called the pseudo-sphere, a world of two unending trumpets.

157

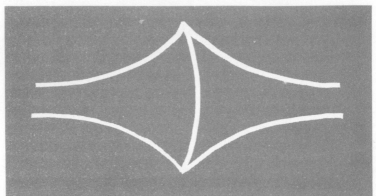

Let us now compare in a few simple respects the "truth" of certain geometrical statements in relation to the plane, the sphere and this pseudosphere. Straight lines, which are "straight" on the plane, follow the surface and therefore curve out on the sphere, curve in on the pseudosphere. Triangles on the sphere curve out; on the pseudosphere, in; and circles appear, depending on the surface, like saucers or limp watches. What happens to geometric "truths"? They are no longer true-false statements, but multiple-choice. The sum of the angles of a triangle is (equal to, more than, less than) 180°. The circumference of a circle is (equal to, more than, less than) $2\pi r$. Through a point not on a given line (one, none, infinitely many) lines can be drawn which will never meet the given line. Which is "true"?

When we compare the geometries of these three very different surfaces, we see that the geometry of one surface cannot be applied to another. We see also that of these three, the surface of the sphere is the one which we can say with greatest accuracy "exists" for us. Yet portions, if not too large, of the imperfect sphere on which we live are more like portions of the Euclidean plane. On the Pacific Ocean we might choose the geometry of the sphere, but in our own backyard we'll take Euclid. So far no one in everyday life has found the geometry of the pseudo-sphere indispensable; nevertheless, logically it is one with the others.

It is interesting to note at this point that if we did not know the nature of the surface of our "earth" we could discover whether the curvature was positive or negative—always provided that it was not too large—in several different ways. Perhaps the simplest would be by adding up the angles of a fairly large triangle. If they added up to definitely more than 180° we would know that we were living on a surface of positive curvative; if to definitely less than 180°, that we had our existence on a surface of negative curvature. But it would be practically impossible to determine with finality that our "earth" was a boundless, endless Euclidean plane. We could never go far enough out so that we could state that the plane was infinite, and we could not even say definitely that it was a plane, or a surface of curvature 0. Whether the total degrees of the three angles of a triangle was exactly 180°, slightly more or slightly less, the range of experimental error would prevent our knowing for sure that it was flat. If, however, our surface is sufficiently large, whether the curvature as a whole is positive, negative or exactly zero, we will find Euclidean geometry most practical because any portion of the surface with which we are concerned will seem, for all practical purposes, flat.

Non-Euclidean geometries were invented not to provide geometries for unusual surfaces but to show that from assumptions other than Euclid's (specifically, a different postulate about parallels) equally *consistent* geometries could be deduced. One of the ways of establishing this consistency is by identifying the objects and relations of Euclidean geometry with certain other objects and relations which result in a non-Euclidean geometry. All of the facts of Euclidean geometry then apply to the model

159

of the non-Euclidean geometry with the exception of the Postulate of the Unique Parallel which, in the case of hyperbolic non-Euclidean geometry, is replaced by the Postulate of Infinitely Many Parallels. It follows, therefore, from the model that the non-Euclidean is as consistent at least as Euclidean geometry.

One of the best-known models of hyperbolic non-Euclidean geometry is that of Felix Klein (1849-1925). In this model the plane of Euclidean geometry is defined as the points of the interior of a circle. Each of these points is defined as a non-Euclidean point, and the chords of the circle are defined as non-Euclidean straight lines. Other definitions are made, but these three will be sufficient to explain the model below where, as we can see, through a given point P not on a straight line AB, infinitely many straight lines can be drawn which will never intersect the given line.

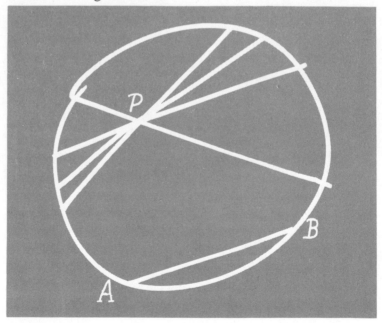

The invention of non-Euclidean geometry freed mathematics from the tyranny of the "obvious," the "self-evident" and the "true," and in so doing served to reveal the nature of mathematics as well as the nature of geometry. With the invention of non-Euclidean geometry, it was recognized for the first time that the theorems of a geometry are logically deduced from a set of arbitrarily chosen assumptions. The truth of the geometry is determined within this framework and has nothing to do with the "truth" (as judged by external facts) of the assumptions from which it is deduced.

We are inclined to think of a geometry as being tailored, as it were, to fit a particular surface; but actually geometries are rather like ready-made suits. They can be used if they fit. Euclidean geometry fits portions of the earth very well, although the idealized type of surface which is implicit in the geometry apparently does not exist at all—"nothing, intricately drawn nowhere." The surface of elliptic non-Euclidean geometry on which we go halfway around and come back to our starting point and the surface of hyperbolic non-Euclidean geometry on which the "ends" of the world become smaller and smaller as they approach infinity are as non-existent as the Euclidean plane. The fact has nothing to do with their mathematical importance. They were not invented to be useful.

It is important that we clearly understand this point, for something happened sometime after their invention which gave to these non-Euclidean geometries the same kind of physical importance that was for so many centuries the unique possession of Euclidean geometry: the geometry for relativity was discovered in a non-Euclidean geometry of boundless, finite, "curved" space. In such a space the geodesics are paths of light waves, which are deflected in varying degrees from their "straight" course by the various masses in space. It is easy to glimpse from

161

just the brief examination we have made of the geodesics of plane, sphere and pseudosphere the implications such deflection would have for any geometry of space.

Mathematically, the usefulness of non-Euclidean geometry was a bonus over and above its mathematical usefulness, which was, as we have seen, the freeing of mathematics from its ancient bonds.

The new freedom, which included freedom from the axioms of Euclid, did not, however, include dispensing with Euclid's axiomatic method. This had been the ideal of all mathematicians since his time. Yet actually it had been hobbled by the definition of an axiom as a self-evident truth. When this definition was dispensed with and an axiom recognized as simply an arbitrary assumption, the axiomatic method became infinitely more valuable to mathematics.

So it is that while Euclid is no longer all geometry, he *is* the axiomatic method—the logical ideal and aim of mathematics and of all science—and the "flaw" which so many generations of mathematicians labored to remove from the work of the master is seen as no flaw at all, but the hallmark of his genius.

11

*Worlds
We Make*

THE COMMON METHOD OF MAPPING BY
coordinates, which enables us to find
our way about an unfamiliar part of
our city, enables mathematicians to
move mentally in a world of n dimensions with as much freedom as they
move physically in a world of three.

The idea of dimensionality has
been with mathematics since the time
of the Greeks. The lengthless breadthless point traced out a line, which had
one dimension. The line traced out a
plane, which then had two dimensions; the plane traced out a space,
which then had three dimensions. No
one with a human desire for consistency could follow this process and fail
to ask the next question. Why not a
figure, a sort of hypersolid, traced out
by a solid moving in a 4-dimensional
space?

But a 4-*dimensional* space! What
could it possibly be like?

Although we cannot visualize a 4-
dimensional space, we can visualize
the effect that "going into the fourth
dimension," as science fiction writers
say, would have on an object from
space of three dimensions. This we can
come to by a logical extension of what
we can actually see of the relation between space and the plane. Let us take
a piece of paper, trace out the soles of
our shoes, and cut them out. We have

163

a right sole and a left sole, mirror images of each other. If we limit ourselves to sliding them around in the plane, on a table top, for instance, we can never make them both left soles. But if we lift the right sole off the table (out of the plane and into space), we can turn it over so that it *is* a left sole when we return it to its mate. Now let us take the shoes, one right and one left, from which we traced the soles. These are 3-dimensional objects in 3-dimensional space. We know from experience that we can never turn the right shoe into a left shoe. But if we were able to lift it out of our space and into a 4-dimensional space, turn it over and return it, *what would have happened to it?*

There is yet another way by which we can get a visual idea of 4-dimensional space. This too is by a logical extension from the three dimensions with which we are familiar. Let us take the simplest figures in each dimension:

A line segment is bounded by two points.
A triangle is bounded by three line segments.
A tetrahedron is bounded by four triangles.

Should there not be, in a 4-dimensional space, a figure bounded by five tetrahedra? This logical extension of the first three figures we call a pentahedroid. When the five tetrahedra are regular, the pentahedroid (it can be proved) is one of the six regular bodies possible in a 4-dimensional space.

What does the pentahedroid look like? Well, it is a figure bounded by five tetrahedra. Although we are somewhat like the Lady of Shalott in that we cannot turn and see it and live, we can look at it in several of the ways in which we usually look at 3-dimensional figures.

We can "see" a hypersolid in a manner similar to the

164

one in which we are accustomed to seeing 3-dimensional solids in two dimensions. We are all familiar with the real appearance of these objects in photographs and paintings. Of course, it is our actual experience with the objects in three dimensions which gives for us a reality to their representation in two, and this actual experience is not possible with 4-dimensional objects. Nevertheless, we can construct a perspective model in three dimensions of a never-seen and never-to-be-seen but *logically thought out* figure in four dimensions. We can give an example which is so simple as to be trivial and yet illustrates exactly the relationship. If we are drawing a tetrahedron and are looking at one of its triangular sides from a position directly in front of it and level and parallel with it, we see only—and draw only—the triangular face of the side of the tetrahedron which is toward us. In this particular case, a triangle is a 2-dimensional representation of the 3-dimensional tetrahedron. In the equivalent 3-dimensional representation of a 4-dimensional pentahedroid we have before us—because our projection of the pentahedroid into our space is "head-on"—a tetrahedron, which is one of the faces of the pentahedroid. In the comparable projection of the tetrahedron, the other three faces have been projected into the three straight lines bounding the triangular face we saw. When we look at the tetrahedron which is the head-on projection of the pentahedroid, each face of the tetrahedron which we see is a projection of a bounding tetrahedron, comparable to the projection of planes into lines in the projection of a tetrahedron into a triangle.

We can also make 3-dimensional patterns of 4-dimensional hypersolids almost as easily as we can make 2-dimensional patterns of 3-dimensional solids. To make

165

a plane pattern of the solid tetrahedron which we have represented below

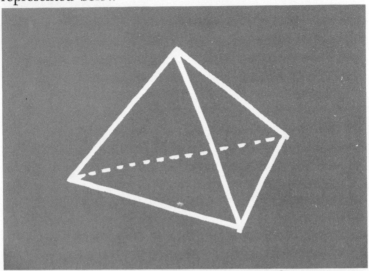

we simply open it up and flatten it out on the page:

An equivalent, but of course 3-dimensional, pattern for a pentahedroid would involve spreading out the hypersolid

166

in space. The resulting pattern would be a tetrahedron
with a tetrahedron upon each face.

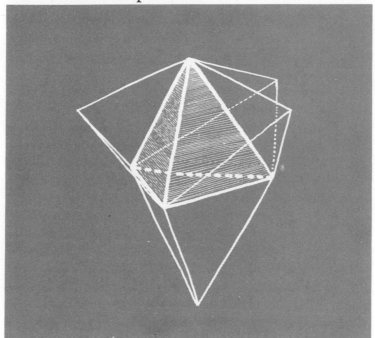

Just as one of our 3-dimensional children would have
no trouble folding the 2-dimensional pattern back into a
3-dimensional tetrahedron, a 4-dimensional child would
make quick work of folding the 3-dimensional pattern
back into a 4-dimensional pentahedroid!

    We can also dismantle a pentahedroid, as if it were a
Tibetan monastery being prepared for shipment to the
home of an American millionaire, the pieces carefully
labeled so that they can be put together again in another
land. Then we should actually have seen a 4-dimensional
body—in pieces!

    It is, of course, impossible to construct an actual
model of a 4-dimensional figure, but mentally we are not

167

so limited. If we do not insist upon an answer to our very human question, "But what does it actually look like?" we can think freely of objects in space of any number of dimensions or, as the mathematicians say, n-dimensional space where n is any number greater than 3.

After our excursion through what might be called the sideshows of dimension theory, let us go back to the system of coordinate axes by which we find our way about an unfamiliar part of our city. These can be the same axes by which we map the points, lines, figures and surfaces of 2-dimensional space in analytic geometry. We saw that any point on the plane could be uniquely located by a number pair (x, y); now we see that any point in three dimensions could be uniquely located if only we had a third axis. This, the z-axis, we erect at the origin perpendicular to the plane formed by the x and y-axes. Now instead of two coordinates, x and y, to locate a point, we need a third, z.

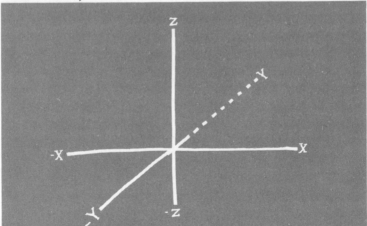

To see how this extension of the system of coordinate mapping works, let us consider the points in the illustration on page 69. On the plane they are uniquely identified by their x and y coordinates as $(2, 1)$, $(-4, 2)$, $(-3, -3)$, $(4, -2)$. If we raise the first two points one unit above the plane and lower the last two points one unit below the plane, we get $(2, 1, 1)$ and $(-4, 2, 1)$ above the plane and $(-3, -3, -1)$ and $(4, -2, -1)$ below the plane. If we raise each of the four original points a different amount, the first one unit, the second two units, and so on, we get the points $(2, 1, 1)$, $(-4, 2, 2)$, $(-3, -3, 3)$ and $(4, -2, 4)$, each one a unique point and each one uniquely identified.

Following the general method we have already outlined, we can locate points, lines, plane figures and solids in 3-dimensional space. The only difference is that instead of expressing these by equations of two variables we shall need equations of three variables. The equation $ax + by + cz + d = 0$ represents a plane in 3-dimensional space just as the equation $ax + by + c = 0$ represents a line in 2-dimensional space.

It is only natural at this point that we ask what is represented by an equation in *four* variables? We do not have to be mathematicians to guess the answer to this one. If an equation of the first degree in two variables represents a line in 2-dimensional space, an equation in three variables represents a plane in 3-dimensional space, then an equation in four variables represents a space (or hyperplane) in 4-dimensional space, and so on.

The reason that we are able to move so freely in n-dimensional space is that, thanks to analytic geometry, we no longer have any need to visualize what we are talking about. We are just talking about algebraic equations. But do not make the mistake of thinking that the geometry of n dimensions is all algebra after $n = 3$. There is a division of labor. Algebra does the work, and geometry suggests the ideas. If, for instance, in 2-dimensional space we have a number pair (x, y) and another pair (x', y'), geometry suggests that we can use in our algebra the concept of "the distance" between (x, y) and (x', y'), since any given number pair can always be represented as a unique point in the plane. The way in which we do this is as old as geometry itself. If we draw a line from (x, y) parallel to the y-axis and a line from (x', y') parallel to the x-axis, the two lines will intersect. When we join (x, y) and (x', y') we have a familiar figure.

By the Pythagorean theorem we know that the hypotenuse of the right triangle, which is also the distance between our two points, is the square root of the sum of the squares of the two sides. We say, then, that our distance formula for two ordered pairs of numbers (x, y) and (x', y') is the one below.*

$$d^2 = (x' - x)^2 + (y - y')^2$$

In the specific case of the two points above $(3, 4)$ and $(7, 1)$, the formula gives us $(7 - 3)^2 + (4 - 1)^2 = 25$.

* Note that the result of squaring $(x - x')$ is the same as the result of squaring $(x' - x)$.

170

The square root of 25 being 5, this is the desired distance between the two points.

When we put our points in a 3-dimensional space as number triples (x, y, z) and (x', y', z'), we have the same formula for the distance between the two points except that we have a third variable, z and z'.

$$d^2 = (x-x')^2 + (y-y')^2 + (z-z')^2$$

We can apply this formula in the following concrete problem. We wish to determine the distance from the back right-hand corner of the top of our desk to the bottom of the front left-hand leg. To do this, we determine first the length of the diagonal of the top of the desk. Then, with this as one side of our right triangle and the front left-hand leg as the other, we determine the length of the hypotenuse, which is the desired distance. Try it some time with a desk.

It is not just the abstract concepts of geometry—like that of distance—which suggest ideas to algebra. Even the geometric figures of space of four dimensions, which we found impossible to visualize a few pages back, become mere formulas and lead us to extensions of themselves in higher and higher dimensions. We are all familiar with the circle and its extension into three dimensions, the sphere. If we map a circle on the plane with its center at the origin, the formula for its radius is

$$x^2 + y^2 = R^2$$

and this means simply that the square of the radius is the sum of the squares of the x and y coordinates of any point on the circumference.

Just as we extended the distance formula into dimensions higher than 2, we can extend the formula for the ra-

171

dius of a circle to the radius of a sphere, a hypersphere, and so on.

$$x^2 + y^2 = R^2$$
$$x^2 + y^2 + z^2 = R^2$$
$$x^2 + y^2 + z^2 + w^2 = R^2$$

We must not think that the mathematics of n dimensions is nothing more than adding another letter for each dimension that we add. Things that are mathematically interesting begin almost as soon as we add that next letter, and they are not at all predictable. If they were, the mathematics of n dimensions might be very useful—which it is— but it would not be very interesting—and it is.

Although the extension of the formula for the radius of the circle into three and four dimensions was made in routine fashion, the extension of the formula for the area of the circle into higher dimensions is not nearly so routine:

For the area of a circle, $\qquad\qquad A = \pi r^2$

For the volume of a sphere, $\qquad\qquad V = \frac{4}{3}\pi r^3$

For the hypervolume of a hypersphere, $H = \frac{1}{2}\pi^2 r^4$

Here we have a very interesting and unexpected relationship. Two generalizations are involved and they alternate, depending upon whether the dimensionality of the figure is even or odd. If the number of dimensions is even, $n = 2k$, we have

$$\frac{\pi^k r^{2k}}{k!}$$

but if the number of dimensions is odd, $n = 2k + 1$, the general expression is quite different.

172

$$\frac{2^{2k+1}k!\pi^k r^{2k+1}}{(2k+1)!}$$

As we go further into the geometry of n dimensions, we find that we never know at just what n our extension may become suddenly more difficult. Consider the problem of packing spheres into space so that in some regular pattern we can fit in the greatest number of spheres. For $n = 2$ we get the most circles on the plane by alternating staggered rows.

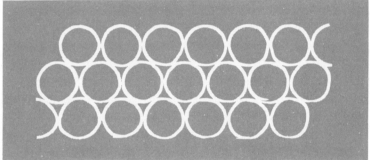

For $n = 3$ we arrange each layer of spheres in the same way that we arranged the circles but stagger the alternate layers. We can continue in similar ways, although it is not at all easy to prove, through $n = 8$. At $n = 9$, the problem inexplicably takes a more difficult turn. At the present time there is no one who can tell us how to pack 9-dimensional spheres in 9-dimensional space!

The geometry of n dimensions might just as well be called the algebra of n variables, but either way the intellectual journey which begins at O on the Cartesian plane takes us through fascinating if purely mental country, and never ends!

173

12

*Where Is In
and
Where Is Out?*

WHEN, AS VERY YOUNG CHILDREN, WE
are told to copy a drawing of a triangle,
we produce a blob. If we are then given
a neat little square to copy, we produce
a brotherly blob. A long thin rectangle
is transformed into a blob, and so is a
circle.

As far as we are concerned, the blob
is a reasonable reproduction of any
number of simple geometric figures. It
is generally admitted that we do not
draw very well; yet we have perceived
the essential likeness of all the figures
we have been given to copy, a likeness
which will escape us in later life when
a rectangle, for instance, will seem like
something entirely different from a
circle.

The fundamental similarity of triangle, square, rectangle and circle is
that they all divide the plane (or the
piece of paper on which they are
drawn) into two distinct and mutually
exclusive parts: that part *A*, which is
inside the boundary, and that part *B*,
which is outside. A point *C* which is in
A cannot simultaneously be in *B*. For *C*
to move from *A* to *B*, it must cross the
boundary of the figure we have drawn,
whether it be triangle, square, rectangle
or circle. If we think of each of these
figures as drawn on a thin sheet of rubber, we can see that no matter how we
pull the sheet about, so long as we do

174

not cut or tear it, we shall never be able to affect in any way this basic and common characteristic.

If, however, we take certain figures like those below which divide the plane, or the paper on which they are drawn, into more than two parts, we shall find that no amount of stretching will turn them into the figures we were first concerned with.

Yet, although we cannot reduce any of these figures to our first simple blobs, we can reduce each of them to a blob with a blob cut out of it; and this is the way, as children, we would have drawn any one of them.

Recalling the straightedge and compass of Euclid's geometry, the protractor in its envelope at the back of the text, the painstaking care with which we drew each figure and lettered the appropriate points, we find it hard to believe that this casual approach to figures can be geometry too. Yet it is. Topology, as this geometry is called, is one

175

of the newest, the most all-inclusive and the most abstruse branches of mathematics. It concerns itself with the truly fundamental properties of geometrical figures, surfaces and spaces. Most of its problems are so removed from our everyday experience that it is impossible for us even to glimpse them, let alone grasp them; yet, as in the higher arithmetic, some of its most difficult problems can be stated in the language of a child.

This is not as surprising as it might at first seem. In an article entitled "How Children Form Mathematical Concepts" (*Scientific American*), Jean Piaget has written:

> A child's order of development in geometry seems to reverse the order of historical discovery. Scientific geometry began with the Euclidean system (concerned with figures, angles and so on), developed in the 17th century the so-called projective geometry (dealing with problems of perspective), and finally came in the 19th century to topology (describing spatial relationships in a general qualitative way—for instance, the distinction between open and closed structures, interiority and exteriority, proximity and separation). A child begins with the last: his first geometrical discoveries are topological. At the age of three he readily distinguishes between open and closed figures: if you ask him to copy a square or a triangle, he draws a closed circle; he draws a cross with two separate lines. If you show him a drawing of a large circle with a small circle inside, he is quite capable of reproducing this relationship, and he can also draw a small circle outside or attached to the edge of the large one. All this he can do before he can draw a rectangle . . . Not until a considerable time after he has mastered topological relationships does he begin

to develop his notions of Euclidean and projective geometry. Then he builds those simultaneously.

Yet the only formal geometries with which most adults are familiar are these last two!

In that with which we are most familiar—the Euclidean geometry we were taught in high school—we studied and proved statements which established the likenesses among different types of figures—triangles, for instance. We were especially fond of the right triangle. Following in the footsteps of Pythagoras, we found that the square constructed on the hypotenuse of the right triangle was equal to the sum of the squares on the other two sides and that all right triangles, regardless of their sizes and shapes, were alike in this respect. (We have seen how this ancient theorem runs through all mathematics: arithmetic, algebra and analysis as well as geometry—we even meet it, in a modified form, in the mathematics of relativity; but one place we never meet it is in topology!)

The other geometry with which we may also have become familiar in high school, in the art course, since it is not taught as mathematics at that level, is projective geometry. (It was Cayley who exclaimed, "Projective geometry is all geometry!"—but it is not topology.) Here, when we attempted to draw the corner of a room, we discovered a curious thing. The corner was formed by the meeting of three right angles and we knew by Euclidean geometry that a right angle is 90° and that the sum of three right angles must be 270°; but when we drew the corner on paper, so that it looked to the eye exactly like the corner we saw, the sum of the three right angles was always 360°! *

* This, of course, is because the corner when projected to the plane on which we are drawing it must fill an entire circle, or 360°.

177

Invariants under rigid motion—length, angle, area—
are the subject of Euclidean geometry. Invariants under
projection—point, line, incidence, cross-ratio—are the subject of projective geometry. (Rigid motions are technically
a class of projections.) No matter how we slide a right
triangle about on the plane, we never affect its "triangle-
ness" nor its "rightness"; but when we draw it from differ-
ing points of view, although we retain its "triangleness,"
we lose its "rightness." The transformations of topology,
which include rigid motions and projections as special
classes, are in general much more drastic. Under the particular group known as the deformations, a right triangle
can be transformed into any other type of triangle, a polygon of any number of sides more than three, an ellipse, a
circle and so on. Yet, through all these changes the characteristic which we perceived when we drew our first
triangle as a blob will remain invariant: it will divide the
plane into two distinct and mutually exclusive parts, an
inside and an outside. This characteristic is invariant
under deformation for any figure like the triangle which
topologists classify as a *simple closed curve*.

Although intuitively we have an idea of what we mean
by a simple closed curve, let us arm ourselves with a more
precise definition. When we think of a curve we probably
think of something the opposite of sharp, angular, straight;
but in mathematics the sharp, the angular and the straight
may all be curves. The ancient definition of a curve is that
it is the path traced by a moving point. In the spirit of this
definition, a closed curve is one whose end point is the
same as its beginning point; and a simple curve is one
which does not pass through the same point more than
once. It is obvious from this definition that circles, triangles, rectangles and higher polygons, as well as blobs,

178

are all simple closed curves. It is not quite so obvious that the figure below is a simple closed curve.

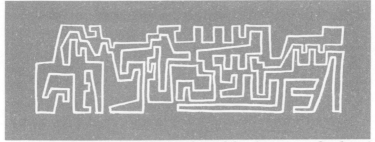

What we perceived so early in life about simple closed curves—that they divide the piece of paper on which they are drawn into an inside and an outside—is one of the fundamental theorems of topology.

THEOREM: *A simple closed curve in the plane divides the plane into exactly two domains.*

There are many mathematical theorems which, in the course of this book, we will receive with puzzled frowns or raised eyebrows; but the Jordan Curve Theorem, as the above is known, is not one of them. This theorem was first stated by Camille Jordan (1838-1922). Besides being a mathematician of the first order, Jordan was a great teacher and the author of a textbook, *Cours d'analyse,* which is an acknowledged masterpiece. In *A Mathematician's Apology* Hardy has stated his own debt to Jordan and to his book as follows: "I shall never forget the astonishment with which I read that remarkable work, the first inspiration for so many mathematicians of my generation, and learnt for the first time as I read it what mathematics really meant. From that time onwards I was in my way a real mathematician, with sound mathematical ambitions and a genuine passion for mathematics."

179

We have included this testimonial from Hardy to make clear that Jordan was a mathematician of stature and influence. If Jordan was interested in the fact that a simple closed curve divides the plane into two domains, it must be more interesting and less obvious than the observation of a three-year-old would lead us to believe. (Actually modern mathematicians have a considerable respect for the obvious. They have found that quite often what appears obvious is not at all; in fact, quite often it is not even true. They have also found that even when it is true, it is often almost impossible to prove that it is true.) Jordan experienced considerable difficulty in trying to prove the obvious theorem which bears his name, so much difficulty that his proof did not meet the rigorous standards which he himself had set up in his *Cours d'analyse*. Time and effort on the part of other mathematicians finally filled the logical gaps in his reasoning. When at last it was completely acceptable from the rigorous point of view, the proof of this "obvious" theorem was nothing for children. It was so extremely technical that even mathematicians found difficulty in following it.

Why should it be so difficult to prove what we have shown is readily apparent even to a three-year-old?

The answer to this question lies in the complete generality of Jordan's theorem. It is simple (relatively) to prove it for any special case of curve. For instance, we can give a simple method for determining whether a given point is inside or outside the labyrinthine "simple closed curve" that we drew on page 179. Incidentally, the reader can first determine that this is, indeed, a simple closed curve by tracing it. He will find that without lifting his pencil and without crossing a line he can go around the entire curve and return to his starting point. It is a little harder to determine whether a given point is inside or outside. To do so, we take a direction which is not parallel to any side of the figure. Although sometimes difficult, this is not impossible, since any straight-edged closed curve has only a finite number of sides and hence of directions. To determine whether a given point is inside or outside the curve, we direct a "ray" in the chosen direction from the point and past the curve. If the ray crosses the boundary an even number of times, the point is outside; if an odd number of times, inside. Below we have applied this method to a fairly simple figure, but the reader should also apply it to the figure on page 179.

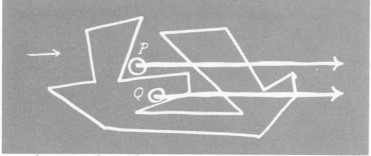

The general problem—in other words, the proof in respect to all simple closed curves—presents difficulties which do not occur in the special case of straight-edged closed curves. *All* simple closed curves include, in addition to the various examples we have already mentioned, such curiosities as curves which have area, curves to which no tangent can be drawn, curves which cross and recross a straight line infinitely many times within an arbitrarily small distance. Although these are contrary to all we think we know about curves, they too may be simple closed curves; and when we make a statement about simple closed curves, as we do in the Jordan Curve Theorem, we

181

are making a statement which must be shown to apply also to such curious curves!

The greatest difficulty of all in proving this theorem is one which seems at first preposterous. *Where is in and where is out?* It is very easy to show that there exists at least one point which is outside the curve. Knowing that the plane is infinite in extent, we select a point sufficiently far away from the boundary so that it is unquestionably outside. But how do we go about showing that there is at least one point which is inside the curve? In the case of the ordinary everyday simple closed curve, the kind which makes the Jordan Curve Theorem seem so obvious, we find our inside point by selecting one which is on the other side of, and an arbitrarily small distance from, the boundary. Even mathematicians agree that such a point is inside. But this method will not be of any use to us when, in going even an arbitrarily small distance across the boundary we shall have already crossed and recrossed the curve an infinite number of times. Such problems, not obvious at all, made the general Jordan Curve Theorem so difficult to prove. Today, proved at last with full rigor and generality for all possible simple closed curves in the plane, the theorem has been extended for their equivalents in space. These are the simple closed surfaces like the sphere and the polyhedra which divide space into two distinct and mutually exclusive parts, that which is inside them and that which is outside.

We again imagine these figures to be made of rubber, thin enough to be stretched at will into any topologically equivalent shape we choose yet strong enough to hold a shape. As we pull them about, what other characteristics about them remain invariant? No matter how we stretch these surfaces, we cannot change the fact that each has two sides, an inner side and an outer side. We also cannot

182

change the fact that they have no edge. These, like the characteristic of dividing space into two parts, are invariant.

If we puncture our general balloon-like surface and carefully stretch it out flat, we get a surface which we can call a disk. This disk, which we can say is the topological equivalent of a sphere with one hole in it, does not of course divide space into two parts because it encloses no space. It is not unbounded, as the sphere is, and therefore it has an edge where the sphere has none. It has, however, one characteristic of the sphere. It has two sides. Unless we are somewhat informed on the subject of topological curiosities, we may think that all surfaces have two sides, and if this is all the sphere and the disk have in common, it isn't much. However, although it is impossible to have a three-sided surface, it is perfectly possible to have a surface with only one side.

We can take our disk, with its two sides and its one edge, and stretch it out into a long thin strip like the one below.

Let us paint one side of this strip red, and one side green. Then let us pick it up and join the two ends so that red meets red and green, green. We have a band which is red on one side and green on the other. Like the strip (from the disk) with which it was formed, it has two sides; but unlike the strip, it has not one edge but two. The original strip was the topological equivalent of a sphere with one hole in it; the band is the topological equivalent of a sphere with two holes in it.

Now let us take another similar but unpainted strip

183

and give it a half twist before we join the ends together. We do not have a band, but something quite different topologically. Where the band has two sides and two edges, the Möbius strip, as it is called, after A. F. Möbius (1790-1868), has only one side and one edge. If we attempt to paint one side red, we shall never find a place to stop until we get back to where we started, and by then the entire strip will have been painted red!

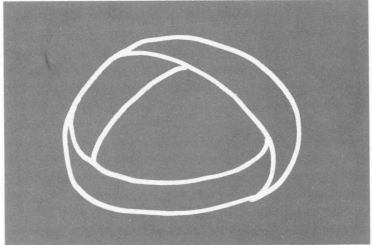

The Möbius strip and the band were both made from a strip which was a stretched-out disk; yet no amount of stretching will enable us to make a Möbius strip into a band or a band into a Möbius strip. What happens, though, when we perform a similar operation upon all three? We cut them down the length. The original strip falls into two strips; the band falls into two bands; but the Möbius strip remains in a single piece! (Try cutting it again.)

Topologists, it is clear, look at things and see them differently from the way most of us do. Where we see a circle or a triangle or a square, a topologist sees a simple closed curve; where we see a knot—just a knot—a topologist sees many different kinds of knots, and he is fascinated by them.

By a knot, a topologist means nothing so simple as even the most complicated knot that Boy Scout, first-aid instructor, cowboy or sailor can tie. A knot which is tied can be untied. It is, therefore, topologically equivalent to the piece of string or rope out of which it was tied, a line segment or a simple open curve. A topologist is interested in knots which are not tied and therefore cannot be untied. Such knots are essentially loops or circles, simple closed curves in space, but with a difference.

The most famous of these is probably the trefoil, or clover leaf, knot pictured below in two different forms.

No amount of stretching or pulling or clever weaving can transform one of these knots into the other. Yet both (in fact, *any* knot) can be mapped upon a simple closed curve—a rubber band, for instance. We put the band and the string out of which the knot is made together at one point and then keep them together at each point as we move around the rubber band. Eventually we come back to where we started, never having had to separate string and rubber band at any point. In this respect a knot is

185

equivalent to a simple closed curve; yet no amount of stretching, nothing short of cutting the knot and rejoining the ends, can make a knot into a simple closed curve, for it is embedded in three-dimensional space in a different way.

We may think of knots only as pleasant puzzles, yet they present topology with one of its greatest unsolved problems: that of classifying different kinds of knots according to their invariants. One method which works very well for the great majority of knots is that of associating each one with a certain surface, the edges of which can be arranged so that they trace out that particular knot. A Möbius strip with three half-twists instead of the usual one, for instance, will trace out in the path of its edge a

trefoil knot. But a *general* method of classification which would cover all cases has not yet been discovered.

Here, as in the proving of the Jordan Curve Theorem, the difficulty lies in the complete generality of the problem; yet if a general method of classifying knots can be found, much in related topological fields will fall automatically into place, like minor candidates riding into office on the leading candidate's coattails.

Perhaps the solution to this problem lies within the

future grasp of some chubby hand drawing circles and triangles as indistinguishable blobs.

"One thing seems certain," wrote E. T. Bell, in *The Development of Mathematics:* "to think topologically, the thinker must begin young. The cradle with its enchained teething rings may be a little too early; but the education of a prospective topologist should not in any case be deferred beyond the third year. Chinese and Japanese puzzles of the most exasperating kind, also the most devilish meshes of intertwisted wires to be taken apart without a single false move, should be the only toys allowed after the young topologist has learned to walk."

Topology is one of the youngest branches of an ancient subject, and much of its strength has come from the youthfulness with which it has looked at age-old figures. It has seen what was always there but never seen before —by grownups.

13

What a
Geometry IS

IF EUCLID WERE TO RETURN TODAY, seeking news of what he loved best, he might be surprised to find in the schools only his own geometry—his own theorems. He might well wonder if nothing at all had happened to mathematics in some twenty centuries. But, in the universities, he would find out what had happened. He would be confronted by not one but many, many different geometries, of which his own was only the most elementary. "A geometry," he would be told in strangely unfamiliar terms, "is the study of those properties of figures which remain invariant under a given group of transformations."

Invariant.

Group.

Transformation.

Even in Greek, these words would have no mathematical meaning for Euclid. Yet with the help of the concepts which they represent, mathematics has been able to bring together into one unified whole all the very different geometries which have been developed in the twenty-three hundred years since Euclid composed his *Elements.*

The key word of the three is "group," a concept which has been called the unifying principle of modern science. For simplicity's sake, how-

188

ever, we shall begin our examination of "what a geometry *is*" with the two less complicated concepts of "invariance" and "transformation," which we have already met in several earlier chapters.

These two ideas are in a sense diametrically opposed. The concept of transformation represents *change:* invariance represents *changelessness.* When we combine the two, we are concerned with that which is *changeless under change.*

Let us take a very simple geometrical example, remembering as we do so that these concepts can be applied to much more than geometry, to much more in fact than mathematics. We pick up a right triangle (A) and move

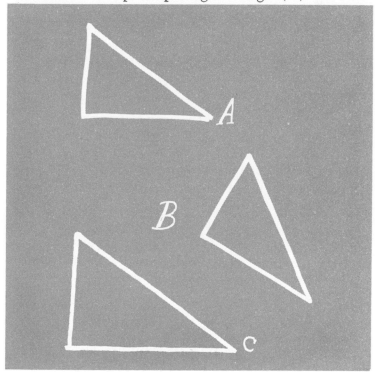

it by what geometers call "rigid motion" from one place to another (B). We find that certain of its geometric properties change but others do not. Its position changes, for instance, but its size and shape do not. If, however, we proceed to expand it in a uniform manner (C), its shape does not change but its size does. We now move it in space so that it is "in perspective" (D) with the position it originally held. It remains a triangle, but it is no longer a right triangle. Now we drape it over a globe (E), allowing the sides to fall along the shortest distances between the vertices. We find that we still have a three-sided figure, but the sides are not straight lines and the angles, unlike the angles of A, B, C and D, add up to more than $180°$. We take up our triangle and s-t-r-e-t-c-h it out between our fingers (F). It remains, like our original triangle, a simple closed curve, dividing the surface on which it lies into two distinct parts, that which is outside the curve and that which is inside—but everything else about the original triangle has changed. It is no longer even a triangle!

We have subjected a given right triangle to four different changes, or *transformations,* and in each case at least one of the properties of our original figure has remained *invariant* under that particular kind of transformation. We recognize that each of the transformations has given us a figure characteristic of one of the geometries which we have already examined in this book: Euclidean geometry, projective geometry, elliptic non-Euclidean geometry and topology, or "rubber-sheet" geometry. Yet we have touched on only a few of the more obvious transformations to which we can subject a right triangle. We are reminded—among other things—of reflections, translations, dilations, inversions and rotations. With each

190

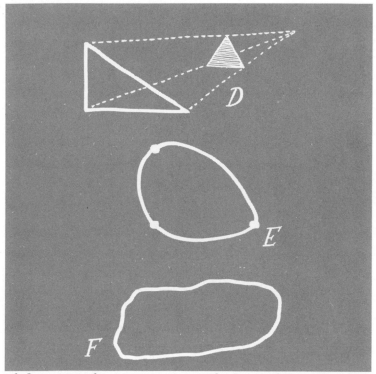

of these transformations we can find ourselves with a different geometry!

Even a hundred years ago, the garden of mathematics seemed rankly overgrown with geometries. Projective geometry threatened to take over the place. "Projective geometry is all geometry!" one enchanted mathematician was heard to exclaim. Yet among the neatly tended rows of Euclidean geometry all sorts of non-Euclidean geometries were springing up. Topology showed a tentative blade of green as *analysis situs*. It was obvious that a period of wild growth now needed to be followed by some attention to pruning.

191

It was at this time that young Felix Klein, whose model of non-Euclidean geometry we have already met in Chapter 10, made a speech at Erlangen University which offered the very tool needed for the pruning. Because of the location where the speech was made, Klein's proposal has come to be known in the history of mathematics as the *Erlangen Program.*

Klein suggested that under an entirely new definition of *geometry,* all of the many apparently disconnected geometries could be brought together, classified and unified. Once more, geometry would be one great subject of study instead of many smaller subjects. The new definition which he proposed is the one we have already met:

"A geometry," Felix Klein suggested, "is the study of those properties of figures which remain invariant under a given group of transformations."

The concept of *a group,* upon which Klein's Erlangen Program depended, had been used earlier in connection with the solvability of algebraic equations. We shall see now how it was used to unify and define the many branches of geometry.

Group is one of those everyday words which a modern mathematician uses in a very precise sense. By it he means nothing so vague as the "assemblage" of Webster. A mathematical *group* must satisfy four specific requirements, which are labeled in the way of mathematics G_1, G_2, G_3, G_4 and are listed on the next page. Without any further explanation, these very abstract requirements would probably seem to the reader entirely removed from the world he knows; yet the group concept is one with which we live and work every day as we use the ordinary operations of arithmetic. The rational numbers, for instance, constitute a group (G) with respect to the operation (o) of addition; and the non-zero rational

192

G_1 If A and B are in G, then AoB is in G.

G_2 If A, B, C are elements of G, the result of operating upon the elements A and BoC, in the order named, is the same as the result of operating upon AoB and C, in the order named, or $Ao(BoC) = (AoB)oC$.

G_3 There exists in G an element I such that $AoI = A$ for every A.

G_4 There exists in G, corresponding to an element A, another A', such that $AoA' = I$ for every A.

numbers, a group with respect to the operation of multiplication. The positive integers, on the other hand, do not constitute a group under either operation.

Let us, therefore, approach the group concept through the positive integers.

We can begin with something as simple as $2 + 2 = 4$.

This we know. All our lives we feel that it is something we can depend upon. It is our symbol for what is changeless in a changing world and, curiously enough, we are not so wrong about *that*. For the fact in which we have such confidence is a specific example of a most general property: a property which has provided mathematics and, through mathematics, the physical sciences with a scalpel for laying bare the very bones of structure.

How is it possible that something as simple as adding together two numbers and obtaining a third of the same kind can lead us to such a unifying concept? To answer this question, we must begin by abstracting from the

193

statement that $2 + 2 = 4$ the general property of which it is a specific example. Let us call our 2's A and B and say nothing more about them other than that they are members of the same class. (As A and B, they may be either the same or different members of the class.) Let us then call the addition represented by $+$ *an operation,* or rule of combination, and designate it o. Instead of $2 + 2 = 4$, we now say that when A and B are members of a class, AoB (the result of combining A and B by the operation o) is also a member of the class. In the same way 2 and 3 are positive integers, and 5—the result of combining them by the operation of addition—is also a positive integer. The property exhibited by A and B in respect to o, and 2 and 3 in respect to $+$, we call *the group property.* We have already met it as G_1, the first of our requirements for a group.

G_1 If A and B are in G, then AoB is in G.

This group property is the first in a succession of abstractions which has made the branch of mathematics called "the theory of groups" something especially abstract even in a subject as abstract as higher mathematics.

To be sure that we thoroughly understand this first abstraction upon which all the others will rest, let us translate it back into the concrete. If instead of A and B, 2 and 3 are members of the class of positive integers; and if we consider in order the common operations of addition, multiplication, subtraction and division, we find that the results of certain operations $(2 + 3)$, or 5, and (2×3), or 6, are also positive integers; but the results of other operations, $(2 - 3)$, or -1, and $(2 \div 3)$, or $\frac{2}{3}$, are not. We say, then, that the positive integers exhibit the group property—or the first requirement of a group— under the operations of addition and multiplication, but not under subtraction and division.

So far we have been using the words *class* and *opera-*

194

tion on the assumption that we know well enough what we mean by them; but before we continue we should do well to pause and define our terms in a more mathematically approved manner. We say that a *class* of objects is defined whenever a rule or condition is given whereby we can tell whether an object belongs or does not belong to the class. If we say "all positive integers," we have defined a class which does not include 0; but when we say "all non-negative integers," we have included 0 in the class which we have defined. We say that an *operation* upon the elements A and B of a class is defined if, corresponding to those elements, there exists a third thing called C, the result. In this general definition of an operation nothing is said about C's being an element of the same class as A and B. When it is an element, as in addition and multiplication of the positive integers—and only then—we can say that the class has the group property under that particular operation.

We have seen that G_1 of the four requirements for a group is merely the statement that a class which constitutes a group must possess the group property. We also recognize G_2 now as the abstraction of the familiar fact that $1 + (2 + 3) = (1 + 2) + 3$ and that $1 \times (2 \times 3) = (1 \times 2) \times 3$, what the textbooks call the Associative Laws of Arithmetic.

G_2 If A, B, C are elements of G, the result of operating upon the elements A and BoC, in the order named, is the same as the result of operating upon AoB and C, in the order named, or $Ao(BoC) = (AoB)oC$.

We already know that among the positive integers the operations of addition and multiplication are associative (although of course subtraction and division are not), so we can move to the third and fourth requirements.

G_3 and G_4 require that in a group there must be two

195

elements which have very specific functions. The first, called the Identity, is an element which when combined with A always gives the result A. The second, called the Inverse, is an element which when combined with A always gives the Identity as the result of combination.

G_3 There exists in G an element I such that $AoI = A$ for every A.

G_4 There exists in G, corresponding to an element A, another A', such that $AoA' = I$ for every A.

Let us consider now whether the positive integers 1, 2, 3, . . . , which exhibit the group property (G_1) and observe the associative requirement (G_2) with respect to addition and multiplication, also meet the third requirement for a group by possessing among their elements both an Identity and an Inverse. G_3 requires for the operation of addition among the positive integers an I such that $A + I = A$. Since 0 is the only number which can be added to an integer without changing its value $(A + 0 = A)$ and since 0 is not included in the class of positive integers, we have to conclude that the positive integers do not constitute a group with respect to addition. In respect to multiplication, however, there is a number, the number 1, by which any integer can be multiplied without changing its value $(A \times 1 = A)$. So, with respect to multiplication, the positive integers do meet the first three requirements for a group.

If they then meet the requirement of G_4, they constitute a group. But G_4 postulates the existence of an element for every member which when multiplied by that member will yield the Identity—in this case, the number 1. There are no such numbers among the positive integers. To meet the requirements of G_4, we must enlarge our class to include the reciprocals of all the positive integers: $\frac{1}{2}$, $\frac{1}{3}$, $\frac{1}{4}$, . . . Then $A \times 1/A = 1$, the Identity.

If, however, we conclude that the positive integers and their reciprocals form a group with respect to multiplication, we shall have fallen into error. Our *enlarged* class no longer exhibits the group property, although our original class of the positive integers did. When we multiply integers and reciprocals, we get results which are neither integers nor reciprocals, and therefore not members of our class:

$$2 \times \frac{1}{3} = \frac{2}{3} \qquad\qquad 3 \times \frac{1}{2} = \frac{3}{2}$$

Doggedly, we enlarge our class once again to include all the positive rational numbers. And now, at last, we have *a group!*

But we have seen that the technical requirements for a group, although they are only four in number, can be slippery things indeed. To discover whether he has them firmly in mind, particularly the requirements for the Identity and the Inverse, the reader should take the simple test at the end of this chapter.

In spite of the fact that there exist infinitely many groups, our chances that a particular class will meet the requirements for a group are relatively slim—just as our chances that a particular number will be a prime are slim, although the number of primes is infinite. For this reason we say that, in spite of the fact that the number of groups is infinite, *almost all* classes with respect to a particular operation are *not* groups.

Up to this point we have been thinking exclusively of groups in which members of a class (like numbers) are combined by a certain operation (like addition or multiplication). We can also think, however, of a group as *a class of operations* which can be performed one after another (the rule of combination, in this case) to yield a result which could have been achieved by a single operation. This is the same as getting "an answer" which is in the class when we combine two members of a class. For example, in the class of whole numbers, the two operations (add —2) and (add +5) when performed in succession yield a result which could have been achieved by the single operation (add +3).

This concept of a group as a class of operations can be better understood when we examine a class of actual physical operations. Consider, for instance, the rotations in the plane which will turn a square, placed with center at the origin, into itself. The members of this class are four in number, the rotations of 0°, 90°, 180° and 270°:

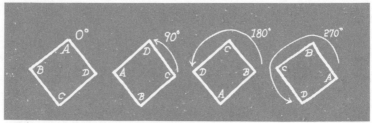

When we subject this class of four rotations to the requirements for a group, where our "operation" is performing one rotation after another, we find that it meets all four requirements, as listed below.

G₁   Any two rotations when performed in succession are the equivalent of performing just one rotation:

     The rotation of 90°, for example, followed by the rotation of 180° is the equivalent of the single rotation of 270°.

G_2 The order of combination of the rotations does not affect the result.

G_3 There is an Identity element—the rotation of 0°—which does not change the effect of any rotation with which it is combined.

G_4 There is for each rotation another, an Inverse element, which when combined with it returns the square to the starting point and is the equivalent of a rotation of 0°, the Identity:

A rotation of 270° followed by a rotation of 90° is the equivalent of a rotation of 0°, since it returns the square to its starting point.

The group of four rotations which will turn a square upon itself is not only a finite group, but a very small finite group. Yet from it we can get a glimpse of the great power of the group concept.

By working out the various possible combinations of our four rotations, we can construct a "multiplication table" for our group, where I, A, B, C are rotations through 0°, 90°, 180° and 270°, respectively:

$$IA = A \qquad AB = C \qquad I^2 = I$$
$$IB = B \qquad AC = I \qquad A^2 = B$$
$$IC = C \qquad BC = A \qquad B^2 = I$$
$$C^2 = B$$

or

	I	A	B	C
I	I	A	B	C
A	A	B	C	I
B	B	C	I	A
C	C	I	A	B

This same multiplication table will work for other groups which do not, at first glance, appear to have any connection whatsoever with the four rotations in the plane

199

which turn a square at the origin upon itself. If, for instance, we take the numbers 1, i, —1 and —i and label them in order I, A, B, C, we shall find that their multiplication table is the same as that of the four rotations:

$IA = A$, or $1 \times i = i$ $AB = C$, or $i \times -1 = -i$
$IB = B$, or $1 \times -1 = -1$ $AC = I$, or $i \times -i = 1$
$IC = C$, or $1 \times -i = -i$ $BC = A$, or $-1 \times -i = i$

$$I^2 = I, \text{ or } 1^2 = 1$$
$$A^2 = B, \text{ or } i^2 = -1$$
$$B^2 = I, \text{ or } (-1)^2 = 1$$
$$C^2 = B, \text{ or } (-i)^2 = -1$$

This should not surprise us when we recall our interpretation in Chapter 7 of the complex number plane as formed by two axes, of the real and imaginary numbers, placed perpendicular to one another. If we concentrate upon that portion of the real axis which is to the right of the origin (the positive reals), we can see that successive rotations of the number plane through 0°, 90°, 180° and 270° are the equivalent of multiplying the positive reals by 1, i, —1 and —i, respectively:

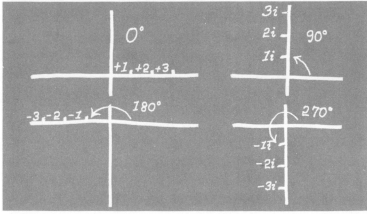

The multiplication table for a group reveals to us what is called its *abstract group*. We have seen that the four rotations in the plane which turn a square into itself and the four roots of unity have the same multiplication table. We know, therefore, that they have the same abstract group, and we can now concentrate upon one group instead of two. What we learn about the abstract group we can apply to the group of four rotations and to the group of four roots of unity as well as to any group of four elements generated by the powers of one element. This means, among other things, that when in the investigation of some phenomenon we come upon the hitherto-unsuspected pattern of our abstract group, the mathematics is already there and waiting for us.

The recognition that several apparently disparate theories have the same abstract group may also result in the discovery of significant and previously undetected relationships among them. Consider the case of a group of rotations somewhat similar to our group of four. This is the group of all those rotations in space which turn a 20-sided regular solid, or icosahedron, upon itself so that after each rotation it occupies the same volume it did before the rotation. The abstract group of these rotations is also the abstract group of certain permutations which we come up against when we attempt to solve the general equation of the fifth degree; the same group occurs in the theory of elliptic functions. The relationship? It turns out that the general equation of the fifth degree, which cannot be solved algebraically, can be solved by means of elliptic functions. Such is the power of the group concept to uncover similarities among apparent dissimilarities!

With the concepts of *invariance* and *transformation* added to the basic concepts of *group* and *abstract group*, mathematics has an unbelievably powerful tool for stripping away the externals and revealing the essentials of structure in the physical world as well as in the mathematical. This tool is not limited in any way. It is a method of looking at any class of any thing under any operation which combines any two members of the class. It is not limited to infinite classes or even to very large classes. It is not limited to classes whose individual members have gaps between them but may be exhibited by classes in which the individual members are, practically speaking, indistinguishable from one another. It is not limited to classes in which all of the elements are essentially the same or in which the same operation is performed upon every pair of elements. We have seen that in mathematics the group concept is not limited to numbers. The idea of groups was first used in connection with the solvability of algebraic equations. Yet it was basic to a program which unified and defined the many branches of geometry.

By utilizing the concepts of invariance, group and transformation, Felix Klein was able in his Erlangen Program to propose a criterion for determining whether a given discipline, perhaps as far removed "from the geometry of Euclid as topology, is "a geometry." Under this great unifying principle we are able to classify some of the varied geometries we have already met in the following manner:

Euclidean geometry is concerned with those properties of geometric figures which are invariant under the group of similarity transformations, while topology is concerned with those properties of geometric figures which are invariant under the group of continuous transformations.

But the group concept, applying equally to algebra and geometry, is not limited even to mathematics. It exhibits itself in the structure of the atom and the structure of the universe. Wherever we can apply the theory of groups, we are able to ignore the bewildering variety, to see among similarities differences and among differences similarities.

The changeless in a changing world!

FOR THE READER

Keeping in mind the four requirements for a group, which are listed on page 193, try to determine which, if any, of the four requirements are met by each of the following classes. Which are groups?

CLASS	OPERATION	G_1	G_2	G_3	G_4
1. All integers	—	—	—	—	—
2. All rationals	+	—	—	—	—
3. All rationals	—	—	—	—	—
4. All even numbers	+	—	—	—	—
5. All even numbers	×	—	—	—	—
6. All odd numbers	×	—	—	—	—
7. 1	×	—	—	—	—
8. 1, —1	+	—	—	—	—
9. 1, —1	×	—	—	—	—
10. 1, —1	÷	—	—	—	—
11. 1, 0, —1	×	—	—	—	—
12. 1, i, —1, $-i$	×	—	—	—	—

ANSWERS

1. $C_{1,3,4}$
2. A group
3. $C_{1,3,4}$
4. A group
5. $C_{1,2}$
6. $C_{1,2,3}$
7. A group
8. C_2
9. A group
10. A group
11. $C_{1,2,3}$
12. A group

203

14

*Counting
the Infinite*

THE INFINITE—PROBLEM . . . PARADOX
. . . and paradise—has been with mathematics since its beginnings. It lies, unstated, in the assumption upon which Euclid's geometry rests. It is implicit in the first numbers with which we begin to count.

0, 1, 2, 3, . . . *

The three dots after these first few numbers indicate to us that they are enough for counting: that we shall never run out of numbers to count with, for there is no last number. The counting numbers are infinite. They are also enough to *count* the infinite, provided it is not too large. They are not, however, enough to count the points on any line, no matter how short!

Before we can understand these paradoxical statements about *counting the infinite,* we shall have to revise our ideas about several things: about "counting," for one, and about "the infinite," for another.

It is quite possible to count without 0, 1, 2, 3, ... A bird that can tell when one of four eggs has been removed from her nest probably has a

---

* The reader may find it difficult to accept 0 as one of the counting numbers, but with what other number will he "count" the unicorns in his living room?

mental picture of the eggs in the nest with which she can "count" the eggs upon her return. Man's first numbers apparently consisted of such grouping pictures—man himself, bird wings, clover leaves, legs of a beast, fingers on his hand—with which other groups could be compared and "counted." If there were as many birds as fingers on his hand, and as many arrows as fingers, then he knew there were "as many" birds as arrows, and an arrow for every bird.

Formally we call what he was doing "counting by one-to-one correspondence" and we probably think of it as a rather inferior trick compared to counting with numbers. Yet what we are doing with our numbers is essentially the same thing. Say that we have a bowl of apples and a party of children. We count the apples and find that we have 7; then we count the children and find that we have 7. We have the same number of apples and children, so we have an apple for every child. We could also have handed an apple to each child and when we came out even we would have known, without knowing the number of children and apples, that we had "as many" apples as children. When we diagram what we have done, we see that in both cases we were counting by one-to-one correspondence very much like man with arrows and birds.

apple ⟷ child	apple ⟷ 1 ⟷ child
apple ⟷ child	apple ⟷ 2 ⟷ child
apple ⟷ child	apple ⟷ 3 ⟷ child *

Counting by one-to-one correspondence is the most

* We have followed here the conventional method of beginning to count with 1; but 0 is logically one of the counting numbers and we can count just as well by beginning with 0. When we do, the answer to the question "How many?" is the successor of the last number which we paired with the last member of the collection.

primitive and also, as we shall see, the most sophisticated method of counting.

The ancient method of directly comparing two collections to determine the number of members is the logical basis for a definition of what we mean by "number" which can be extended to infinite as well as finite collections. Let us firmly banish 0, 1, 2, 3, . . . from our minds for a moment and think instead of all the finite collections we might possibly want to "count" being grouped in such a way that all those which can be placed in one-to-one correspondence with each other—all the collections of a dozen members, for example—are in the same group. These groups do not need to be arranged in order of the size of their respective collections. For the moment it is sufficient for our purposes that they have been grouped. We have all those collections whose members can be placed in one-to-one correspondence with a dozen eggs, all those whose single member can be placed in one-to-one correspondence with the sun, and so on.

MODEL COLLECTION	COLLECTIONS WHICH CAN BE PLACED IN ONE-TO-ONE CORRESPONDENCE WITH MODEL COLLECTION
Day, Night	eyes, antlers, wings, man and woman, good and evil, . . .
Breakfast, Lunch, Dinner	ears and mouth, clover leaves, man-woman-child, stars in Orion's belt, . . .
Sun	head, self, earth, moon, god, . . .

Now, instead of having to keep in mind the specific collections we are using for our models, we can substitute

206

an *X* for each member so that we have *XX, XXX,* and *X.*
We can then easily arrange these new model collections
in the order of their increasing size and, if we want, can
give them names. We are now ready to define *A*, or whatever name we have given the model collection *X*, as the
cardinal number of any class whose members can be
placed in one-to-one correspondence with *X,* or the Sun.
If someone objects and says that all we have done is to
define the number 1, why we shall be generous and call
A "1." Then we shall call our next largest model collection
"2" and define it as the cardinal number of any class
whose members can be placed in one-to-one correspondence with *XX,* or Day and Night; and so on, *to infinity.*

The number of cardinal numbers we can define in this
way is infinite, but the members of each collection in the
classes so defined will be finite. The number of members
in each collection may be very large: all those collections
whose members can be placed in one-to-one correspondence with all the stars in the Milky Way, all those whose
members can be placed in one-to-one correspondence
with all the grains of sand on the earth, all those whose
members can be placed in one-to-one correspondence
with all the electrons in the universe. It may be personally
impossible for us to count all the members of a particular
model collection, but they are "countable" in the sense in
which we commonly use the word. The cardinal numbers
which we have defined are finite cardinal numbers.

But is there any reason why in this same way we cannot define transfinite cardinal numbers for classes which
contain an infinite number of members?

It is at this point that we must change our idea of
"the infinite." For instance, instead of thinking of the
counting numbers 0, 1, 2, 3, . . . as an ever-growing pile
filling room, world, universe, . . . , we must think of them
stuffed, as it were, into the metaphorical suitcase of their

class. In short, we must think of them not primarily as infinite in number but as an infinite class, something which we can handle as a unit, just as we handle finite classes, but something which is still different from a finite class because of the fact that it is infinite. This was not an easy idea, even for mathematicians, to accept. Yet once we accept it, we have something "capable not only of mathematical formulation, but of definition by number." These are the words of the man who, almost singlehanded, corralled the infinite for mathematics.

Georg Cantor, whom we met before as one of the authors of the Cantor-Dedekind axiom, was one of those rare people who are able to look at the familiar as if they have never seen it before and thus become the first to see it. How revolutionary was his idea of the infinite, as *something consummated,* is shown by his own words in presenting it to his mathematical colleagues: "This conception of the infinite is opposed to traditions which have grown dear to me, and it is much against my own will that I have been forced to accept this view. But many years of scientific speculation and trial point to these conclusions as *a logical necessity.*"

Once we have recognized counting as matching one class to another in one-to-one correspondence and an infinite number as something consummated—an infinite class—we are ready to take the next step, which is counting the infinite by placing one infinite class in one-to-one correspondence with another! Doing this, and even the specific way of doing it, was not original with Georg Cantor, living and creating in nineteenth-century Germany and fighting an abstractly bloody battle not only with his colleagues but also with a mathematical tradition of the infinite which went back to the Greeks.

Three hundred years before Cantor, in the Italy of the Inquisition, Galileo had pointed out that the infinite class of squares can be placed in one-to-one correspondence with the infinite class of natural numbers: that there are fully "as many" squares as there are natural numbers, since every number when multiplied by itself produces a square.

Unfortunately, Galileo, with Cantor's theory of the infinite in his palm three hundred years before Georg Cantor was even born, dismissed it: "So far as I see, we can only infer that the number of squares is infinite and the number of their roots is infinite; neither is the number of squares less than the totality of all numbers, nor the latter greater than the former; and finally the attributes *equal, greater,* and *less* are not applicable to infinite, but only to finite quantities." *

What Georg Cantor did three hundred years after Galileo was to take the attributes of *equal, greater* and *less* and apply them to infinite quantities.

When we take the first few numbers and set them off according to some of the various classifications which have been made, we come out with something like this:

ALL NUMBERS	ODD NUMBERS	ODD PRIMES	$4n + 1$ PRIMES	SQUARE NUMBERS
0	1	3	5	0
1	3	5		1
2	5	7		4
3	7			9
4	9			
5				
6				
7				
8				
9				

* Galileo spoke here through the character of Salviatus in his *Mathematical Discourses and Demonstrations.*

209

If we total these various classifications, we find that among the first ten numbers we have five odd numbers, four squares, three odd primes, and only one prime of the form $4n + 1$. We have no trouble in determining that the class of numbers from 0 through 9 is *greater* than any of these sub-classes, that the odd numbers and the even numbers are *equal*, and that the class of primes of the form $4n + 1$ is *less* than any of the other classes. If we attempt to place any of these sub-classes in one-to-one correspondence with the numbers from 0 through 9, we shall have at least five numbers left over. But what happens if, in following the same system, we take, instead of the first ten, *all* of the natural numbers and *all* of the members of the same sub-classes?

ALL NUMBERS	ALL ODD NUMBERS	ALL ODD PRIMES	ALL $4n + 1$ PRIMES	ALL SQUARE NUMBERS
0	1	3	5	0
1	3	5	13	1
2	5	7	17	4
3	7	11	29	9
4	9	13	37	16
5	11	17	41	25
. . .	. . .	. . .	. . .	. . .

It is already apparent. The three dots at the end of each column indicate that each class of numbers is infinite; in spite of the fact that we appear to be exhausting some of the classes, like the $4n + 1$ primes, more quickly than the others, we only appear to be doing so. We can never exhaust an infinite class. When we consider a finite class of whatever size we please, the natural numbers in the chosen class will far outnumber any one of the sub-classes; but when we take all of them, they are *equal* to any one of the equal sub-classes.

210

Galileo said that they were neither more nor less, and that the attribute of *equal* was not applicable to infinite quantities. Cantor said that infinite quantities are equal when they can be placed in one-to-one correspondence with each other: *they have the same cardinal number!*

Just as we said that all classes which could be placed in one-to-one correspondence with the class of the Sun, or X, had the same cardinal number, which we call 1, Cantor said that all classes which can be placed in one-to-one correspondence with the natural numbers have the same cardinal number, which he called aleph-zero or $\aleph_0$. It is different from the finite cardinals only in that it is *transfinite*.

We have already seen how sub-classes of the class of natural numbers can be placed in one-to-one correspondence with the whole of which they are a part; but so curious are the workings of infinite classes, as opposed to finite classes, that we can also do our pairing the other way around. We can set off in one-to-one correspondence with the natural numbers a class of numbers of which they themselves are a sub-class. The class of all integers has one peculiarity which its sub-class, the natural numbers, does not have: it has neither a last nor a first member. How, then, can we pair it off with the natural numbers? This is not so difficult as it might seem. It is merely a matter of ordering the integers in such a way that they can, as it were, stand up and be counted. With no beginning, we begin right in the middle at 0 and then count each pair of integers, positive and negative, in turn.

$$0 \quad 1 \quad 2 \quad 3 \quad 4 \quad 5 \quad 6 \quad 7 \quad 8 \ldots$$
$$\updownarrow \quad \updownarrow \quad \updownarrow \quad \updownarrow \quad \updownarrow \quad \updownarrow \quad \updownarrow \quad \updownarrow \quad \updownarrow$$
$$0 \; +1 \; -1 \; +2 \; -2 \; +3 \; -3 \; +4 \; -4 \ldots$$

There is no particular trick to pairing the natural numbers with the integers, which include them as a sub-class; but such a pairing does serve to show an important technique in counting the infinite. A class of numbers which may not appear to be countable (in the case of the integers, because there is no first number) can often be rearranged in such a way that it can be counted. Consider the class of all positive rational numbers. These are numbers of the form a/b where a and b are both integers. When a is smaller than b, we have what we called in grammar school a "proper" fraction; when b is smaller, an "improper" one. The class of all positive rational numbers is no straightforward sort of infinity like the class of integer squares where we have just one member of the class for each integer. Just one small sub-class, a/b where a is 1, is infinite in number. Since a may take any integer value and for every a, b may take any integer value, we appear to have among these numbers infinity upon infinity, an infinite number of infinities.

If we take the positive rationals in what might be called their natural order, omitting those with common factors since they are already represented, we find that placing them in one-to-one correspondence with the natural numbers is impossible. Not only is there no "smallest" fraction, but also there is no "next largest" fraction. Between any two a/b and c/d an infinity of fractions larger than a/b and smaller than c/d spring up to vex us. Obviously it is impossible for us to pair off with the natural numbers a class of numbers which behave in this fantastic fashion. We have sown dragon teeth on the number line.

But remember, we have said nothing about the rational numbers having to be paired off in their natural order—only that they must be paired in such a way that we can see that we are going to be able to count them with the natural numbers. So let us rearrange the rational numbers. Let us organize them into battalions: the first battalion consisting of all those rational numbers whose numerator is 1, the second battalion consisting of all those whose numerator is 2; and so on.

$$\tfrac{1}{1} \ \tfrac{1}{2} \ \tfrac{1}{3} \ \tfrac{1}{4} \ \tfrac{1}{5} \ \ldots \ \tfrac{2}{1} \ \tfrac{2}{3} \ \tfrac{2}{5} \ \tfrac{2}{7} \ \ldots \ \tfrac{3}{1} \ \tfrac{3}{2} \ \tfrac{3}{4} \ \ldots$$

This arrangement is reminiscent of one of those parades during which we wait restlessly for the band while an apparently endless procession of foot soldiers goes by. The only difference between our parade and the actual parade is that it is not just seemingly endless; it *is* endless. The band, or even the second battalion, can never pass by. Obviously, again it is impossible to count off by placing in one-to-one correspondence with the natural numbers a set of numbers which behave in this fashion; for although in counting the primes, for instance, we would never finish, we would always be able to count as far as any prime we might care to choose. With this arrangement of the rational numbers, not only could we never get to the end, but we could never get to $\tfrac{2}{3}$! Have we then come at last upon an infinity which is impossible to pair with the natural numbers, an infinity whose cardinal number is different from and perhaps larger than $\aleph_0$?

No, we have not.

The simple method by which Georg Cantor ordered the positive rational numbers so that they can be placed in one-to-one correspondence with the natural numbers has the quality of genius. All he did was to take the groupings which we have called battalions and arrange them in rows instead of in one long straight line.

$\tfrac{1}{1}$	$\tfrac{1}{2}$	$\tfrac{1}{3}$	$\tfrac{1}{4}$	$\tfrac{1}{5}$	$\tfrac{1}{6}$	$\ldots$
$\tfrac{2}{1}$	$\tfrac{2}{3}$	$\tfrac{2}{5}$	$\tfrac{2}{7}$	$\tfrac{2}{9}$	$\tfrac{2}{11}$	$\ldots$
$\tfrac{3}{1}$	$\tfrac{3}{2}$	$\tfrac{3}{4}$	$\tfrac{3}{5}$	$\tfrac{3}{7}$	$\tfrac{3}{8}$	$\ldots$
$\tfrac{4}{1}$	$\tfrac{4}{3}$	$\tfrac{4}{5}$	$\tfrac{4}{7}$	$\tfrac{4}{9}$	$\tfrac{4}{11}$	$\ldots$

At this point we might stop for a moment and see, with

this much of a hint, whether we can now order the rationals in such a way that every one will be paired with a unique natural number and whether we will be able to count with the natural numbers to any rational we choose, such as ⅔. . . .

Cantor's way was to order them diagonally, beginning in the upper left-hand corner with ⅟₁.

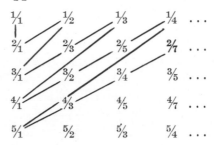

Thus we have all the rationals placed in one-to-one correspondence · with the natural numbers and we quite promptly get to ⅔.

0	1	2	3	4	5	6	7	8	9	...
↕	↕	↕	↕	↕	↕	↕	↕	↕	↕	
⅟₁	²⁄₁	½	³⁄₁	⅔	⅓	⁴⁄₁	³⁄₂	⅖	¼	...

This is mere child's play compared to the task of arranging the algebraic numbers so that they too can be placed in one-to-one correspondence with the natural numbers. The algebraic numbers are all those numbers which are roots of algebraic equations of the form

$$a_0 x^n + a_1 x^{n-1} + \ldots + a_{n-1} x + a_n = 0$$

in which the coefficients $a_0, a_1, \ldots, a_n$ are all integers. This is nothing more than the general expression for the algebraic equations with which we are familiar where n has a value of 1 or 2. When $n = 1$, we have a simple equa-

214

tion like $2x - 1 = 0$, where we can see at a glance that the root, or value of x, must be ½. When $n = 2$, we have a familiar quadratic equation like $3x^2 + 4x + 1 = 0$, where the roots, or values of x, are —1 and —⅓. The essential thing for us to remember is that when such an algebraic equation has whole-number coefficients, as in our examples, it *always* has a root among the complex numbers. (This is the Fundamental Theorem of Algebra, proved by Gauss.) Those complex numbers which can be roots of such algebraic equations are called the *algebraic numbers*. They are not, as we shall see, all of the complex numbers by any means.

Cantor's proof that these algebraic numbers can be placed in one-to-one correspondence with the natural numbers has been called "a triumph of ingenuity"; yet it is essentially as simple as the alphabetization of the telephone book. The crux of the method is what Cantor called the *height* of an algebraic equation. This is the sum of the absolute values of the coefficients plus the degree of the equation less 1. (The absolute values are the numerical values of the coefficients with no attention paid to whether they are positive or negative; the degree is the highest power of the unknown x, or the value for n in the general expression as given above.) Thus the equation of the third degree

$$3x^3 - 4x^2 + 5x - 5 = 0$$

has a height of 19, since $3 + 4 + 5 + 5 + (3 - 1) = 19$.

Having assigned for every algebraic equation a method of determining its height as an integer, Cantor proved that for any integer there is only a finite number of equations which have that particular integer for their height. From this point on, the method of the phone book comes in handy. When we have ordered all algebraic equations according to their height, we find that in most cases we have more than one equation of a particular height. Undaunted, we arrange the equations of the same height according to the value of their first coefficient and, where the first coefficient is the same, according to the second, and so on. Since there is only a finite number of equations with the same height, and since no two equations can have exactly the same coefficients, we have assigned every algebraic equation to a unique position in an order arrangement.

Our purpose, however, is not to order the equations but to order the numbers which can be their roots—the *algebraic numbers*—so that they can be placed in one-to-one correspondence with the natural numbers. So we continue by taking the roots of the ordered equations, which may be more than one but are never more than the degree of the equation, and arranging them according to their increasing value, first according to the value of the real part and then, where several numbers have the same real part, according to the value of the imaginary part. By agreement, as in the case of the rational numbers, we throw out those which are repetitions. We now have a method by which every number which can be the root of an algebraic equation can be paired with one of the natural numbers—this in spite of the fact that we have not actually written down the roots of a single equation!

Cantor's "triumph of ingenuity" can be best appreciated when we recall our diagram of the complex number plane as formed by axes of the pure imaginary and of the real numbers and recall that, although the algebraic numbers are not all the numbers upon the plane, they are everywhere dense upon it, while the natural numbers mark only the units on one-half of the real-number axis!

216

Yet these two seemingly unequal classes have the same cardinal number, $\aleph_0$.

Is $\aleph_0$ the only transfinite cardinal?

We are beginning to suspect that perhaps it is. We have examined many infinite classes of numbers which represent certain specific points upon the complex number plane. All of them are, of course, sub-classes of the complex numbers. Some are sub-classes of the natural numbers as well, and some include the natural numbers as one of their sub-classes. Yet always we have found (with Cantor) that the classes we have examined can be ordered in such a way that they can be placed in one-to-one correspondence with the natural numbers and, therefore, have the same transfinite cardinal, $\aleph_0$.

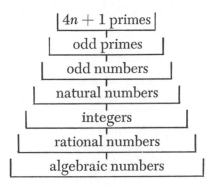

Although we can define infinite classes as being *equal*, it seems that we cannot define them as being *greater* or *less*. Perhaps we were right to begin with: an infinite number is just an infinite number. Fortunately, we were wrong. If we were right, the infinite would be an infinitely less interesting subject than it is. There *is* a transfinite cardinal greater than $\aleph_0$—there is, in fact, an infinite number of greater transfinite cardinals! But at the mo-

217

ment we shall be satisfied with only one. We can find an infinity which is greater than the infinity of natural numbers on a very small part of the real-number line: the segment between 0 and 1.

To show that these real numbers, which are the equivalent of all the points on the segment, cannot be placed in one-to-one correspondence with the natural numbers, Cantor began by *assuming* that they could be. This is a method of mathematical proof as old as Euclid, who used it to show that the number of primes is infinite. It was also used by Fermat to show that all primes of the form $4n + 1$ can be expressed as the sum of two squares. In this case, to prove that the placing of the real numbers in one-to-one correspondence with the natural numbers is impossible, Cantor risked assuming that such a pairing was indeed possible.

As we saw in Chapter 4, all numbers on the real-number line between 0 and 1 can be represented as never-ending decimal fractions, and this is the way in which Cantor chose to represent them. If, however, we start to write down the actual decimals, we immediately become involved in all sorts of difficulties. The first would be 0.000000000 . . . with the 0's continuing to infinity; but what would be the second decimal? No matter how many 0's we place between our decimal point and our first positive place value, we can always construct a smaller decimal by inserting one more 0 and moving our first positive place value over one more place to the right.

0.00000000000000000000000000000000000001 . . .
but
0.000000000000000000000000000000000000001 . . .

Have we proved, then, that it is impossible to arrange the real numbers from 0 to 1 in such a way that they can be placed in one-to-one correspondence with the natural numbers? No. We have proved nothing of the kind. Only that we have not been able to find a way of doing what we want to do. The question then becomes, not whether we can find a way, but whether there *is* a way.

To prove that there *isn't* a way, we begin by assuming that there *is*. We solve the problem of determining the second decimal and all succeeding decimals by assuming that they have been determined. We then think of them abstractly as expressions like $0.a_1a_2a_3a_4a_5a_6a_7$ . . . with each a_n denoting the particular value (0, 1, 2, 3, 4, 5, 6, 7, 8 or 9) of each place in the decimal; and we place them in one-to-one correspondence with the natural numbers, in accordance with our assumption that they can be so placed.

0	$0.a_1a_2a_3a_4a_5a_6a_7a_8a_9$ . . .
1	$0.b_1b_2b_3b_4b_5b_6b_7b_8b_9$ . . .
2	$0.c_1c_2c_3c_4c_5c_6c_7c_8c_9$ . . .
3	$0.d_1d_2d_3d_4d_6d_5d_8d_7d_9$ . . .

. . .   . . .

Cantor showed that such an assumption was false because, even assuming that all decimals could be and had been placed in one-to-one correspondence with the natural numbers, he could construct a decimal which had not been included in the class of "all" decimals so ordered. This decimal he indicated by

$$0.m_1m_2m_3m_4m_5m_6m_7m_8 \ . \ . \ .$$

m_1 being any digit (except 9)* other than the digit rep-

* Since terminating decimals like .25 can be represented as non-terminating decimals in two ways: either as .250000 . . . or as .249999 . . . , we exclude 9 to avoid having our new decimal a different representation of a number which has already, in a different form, been included in the class of "all" decimals.

resented by a_1 in the first decimal; m_2 being any digit (except 9) other than the digit represented by b_2 in the second decimal; and so on. This new decimal would be one not included in the original class of "all" decimals because it would differ from every included decimal in at least one place: from the first in at least its first place, from the second in at least its second place, and so on.

We can see a little more vividly what Cantor did if we take a concrete set of decimals and then by following his method construct a decimal not in our set.

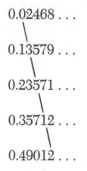

0.02468 . . .

0.13579 . . .

0.23571 . . .

0.35712 . . .

0.49012 . . .

To get a decimal not in the set, we make the first place of our new decimal different from 0; the second, from 3; the third, from 5; and so on. It will differ in at least one place from any decimal in the set: 0.14623 . . . is not included, and there are many other possibilities.

It is almost impossible to overestimate the importance of this achievement. Already Cantor had shown that the attribute *equal* was applicable to infinities; now he showed that the attributes *greater* and *less* were also applicable. The new cardinal number, which is easily shown to be larger than $\aleph_0$, the cardinal number of a "countable" infinity, is $\mathfrak{c}$ (pronounced like "c"), the number of what Cantor called *the continuum*—an "uncountable" infinity!

What other infinities have this same $\mathfrak{c}$ as their cardinal number?

The answer to this question is completely contrary to intuition. We have noted that the real numbers from 0 to 1 are equivalent to the points on the segment of the real number line from 0 to 1, just as all the real numbers are

220

the equivalent of all the points on the line. Our intuition tells us that the infinity of real numbers must be greater than the infinity of real numbers between 0 and 1, just as the infinity of points on the line must be greater than the infinity of points on the line segment between 0 and 1. Yet it is very easy to prove that for every point on the long line there is a point on the short line and that, therefore, there are as many real numbers between 0 and 1 as there are in all the length of the real-number line!

To prove this statement, we shall take two lines (one short, which we shall call *AB,* and one somewhat longer, which we shall call *CD*) and place them parallel to one another. We shall then construct one line which passes through *A* and *C* and another line which passes through

B and *D*. The intersection of these two lines we shall call *O*. It is clear that we can draw a line from *O* to any point *Q* which we choose on line *CD*, and that this line *OQ* will of necessity intersect line *AB* at some point *P*.

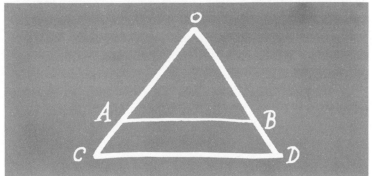

For every Q on the longer line there will be a unique point P on the shorter line which can be placed in one-to-one correspondence with it.

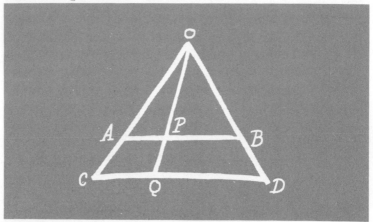

It is also possible to prove, although not so easily, that all the points on the plane can be placed in one-to-one correspondence with the points on a line segment of any finite length. All of these infinities of points have the same cardinal number, c. Since the real numbers represent all the points on the line, and the complex numbers all the points on the plane, they also have c as their cardinal number.

Now that we have distinguished between two types of infinities, those which, like the natural numbers, are "countable" and those which, like the real numbers, are "uncountable," we might think that we were finished with the subject of the infinite. But the infinite is not so easily disposed of.

There are an infinite number of transfinite cardinals which are greater than c, which is greater than $\aleph_0$.

This important fact in the arithmetic of the infinite is stated by a very simple theorem to the effect that

$$2^n \text{ is always greater than } n$$

and supported by a very simple proof. If we consider this theorem when n is a finite cardinal number, we can see that it is true. We take n blocks—n in this case being equal to 3—and paint each block either blue or red. The number of possible color schemes will equal 2^n, or $2^3 = 8$ in this case.

1	2	3
B	B	B
B	B	R
B	R	B
B	R	R
R	B	B
R	B	R
R	R	B
R	R	R

As here, when n is a finite cardinal, we can actually count the color schemes and can actually see that we have exhausted the possibilities: no one can turn up with another color scheme for three blocks painted either red or blue which is not already among the color schemes we have.

But now let us take $n = \aleph_0$. Let us take as many blocks as we have positive integers. Again, let us paint each block either blue or red. How many possible color schemes can we have? Certainly an infinite number. For instance, we could in each case paint the nth block blue and all the others red.

    1.   First block blue and all others red.
    2.   Second block blue and all others red.
    3.   Third block blue and all others red.

   . . .

Obviously this is too easy. After we have paired a

unique color scheme with every one of the positive integers, we can think up one or an infinity more of schemes which we have not included. For instance, we could paint the nth and the $(n + 1)$th block blue and all the others red, and this would give us a completely different set of color schemes which could also be placed in one-to-one correspondence with the positive integers. But remember that even another infinity of color schemes does not prove that both sets of color schemes could not be placed in one-to-one correspondence with the positive integers, or even all possible color schemes!

So let us assume that by some method we have determined all possible color schemes and to each block we have attached one of the color schemes. *Now* can we come up with a color scheme which is not among those attached to the blocks? We can and we do—using the same method by which we constructed a decimal which was not in our original set of "all" decimals. We pick up the first block and note from the color scheme attached to it what color it is to be painted in that particular scheme. Then we paint it a different color, red if it was blue on the list, blue if it was red. The color scheme which results from our newly painted blocks cannot possibly be one of those already attached—or paired in one-to-one correspondence—to the blocks. It will differ, for at least one block, from each of the color schemes we already have. The cardinal number, then, of all possible color schemes is greater than the cardinal number of the blocks because the color schemes cannot be placed in one-to-one correspondence with the blocks. Our theorem—2^n is greater than n—is true whether n is finite or transfinite.

It follows, therefore, that for any transfinite number there is always another and greater transfinite number.

There is no *last* transfinite number. The number of transfinite cardinals is infinite!

Of this infinitude of transfinite numbers $\aleph_0$, as its subscript indicates, is the first. What is $\aleph_1$? The cardinal number of the continuum, c, is larger than $\aleph_0$. There is no known transfinite number that is smaller than c and larger than $\aleph_0$. But is c the second transfinite number? Is it $\aleph_1$? *

In modern mathematics this problem holds the place that the problem of the trisection of the angle held in ancient mathematics. We have indeed counted the infinite, *but we are not done with it!*

* That c is $\aleph_1$ is the famous "continuum hypothesis."

15

*A Most
Ingenious
Paradox*

SEGMENTS OF LINES HAVE LENGTH. SUR-
faces have area. Solids have volume.
The measure assigned to a figure—
length, area or volume, as the case may
be—is unaffected by rigid motion of the
figure. The whole is greater than any
part, and is the sum of all the parts
together.

These statements are as ancient as
Euclid and at the same time so commonplace that we cannot conceive of
their being controverted. Yet in the
theory of point sets, a branch of mathematics in which the paradoxes are almost as numerous as the points (and
the points are very numerous indeed),
we are forced to the conclusion that
under certain conditions, involving the
most familiar figures of geometry, some
of the statements we have made are
"untrue."

To understand the necessity for
this conclusion, we must go back to
that unfortunate Pythagorean who discovered that there can be no rational
number for the point on the measuring
stick which coincides with the diagonal
of the unit square, and perished at sea
for his pains. From this point, quite
literally, we are logically committed to
the theory of point sets, although the
theory itself was not founded until
some twenty-five hundred years later.
When, toward the end of this chapter,

226

we find ourselves balking at some of the conclusions at which we arrive, we must remember that here at the beginning we easily accept—in fact, insist upon—the assumption from which the conclusions will necessarily follow. Who among us would now renounce the idea that for every length there is such a unique measure as $\sqrt{2}$ for the diagonal of the unit square?

The logical consequences of this concept of a number for every point on the line, or the theory of point sets, will be the subject of this chapter. In the course of it we shall find ourselves juggling infinities and distinguishing precisely between those which are non-denumerable and those which are denumerable; transforming by rigid motion whole infinities of points; selecting single points from infinities. Unfortunately, this is not material that can be skimmed. We can only remind the reader that there is no royal road to even the faintest understanding of the concept of point sets, and assure him that if he follows the rocky road of reasoning he may be more than repaid by the satisfaction he gets from a personal contact with pure mathematics.

We must begin by considering what we mean by "a point." When we take a pencil and make with it on paper what we call a point, we have what for all practical purposes is a point. But a point (mathematicians agreed about the time of the Pythagorean) is that which has position but no magnitude. Since any representation of a point must have magnitude, it cannot be a point. More recently, since the time of Descartes, mathematicians have based their definition of a point on its representation by numerical coordinates. A point on the line is a real number. A point in the plane they define as an ordered pair of real numbers; a point in space, as an ordered triple of real numbers; and so on. It is from this definition of a point as a number, and a number as a point, that the great paradoxes of point-set theory develop.

When we start to think of points as numbers, we gain an advantage in handling them. Each one becomes an individual, easily distinguishable from all the others. We can divide an infinity of points into mutually exclusive sets and have no trouble at all in determining whether a given point belongs in a set. All the points on the line, for instance, can be divided into those which represent a real number less than 0 and those which represent a real number greater than 0; while a third set, the single point 0, serves as the boundary between the other two sets.

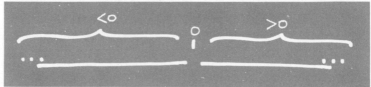

We can make a similar division of the points on the plane by including in one set all those the x-coordinate of which is less than 0 and in the other, greater than 0. Here the boundary set will contain not just one point but all those points with $x = 0$, or the y-axis itself.

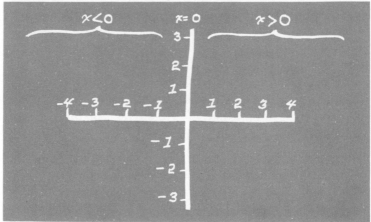

If we inscribe a figure on the plane—let us say a circle of radius 1 about the origin—we can distinguish the points which are on its circumference from all the other points in the plane. Physically, this is impossible; for our drawing, no matter how finely done, must add magnitude to the position of the points. Mentally, though, such a selection is perfectly possible.

The equation for the given circle is

$$x^2 + y^2 = 1$$

since, by the Pythagorean theorem, the sum of the squares of the x and y coordinates at any point on the circumference will give us the square of the hypotenuse, which is also the square of the radius of the circle, in this case 1. There are various sets of points which we can represent by means of this knowledge. The equation itself is the equivalent of the statement "all the points x, y for which the equation holds." If we take at random two points, say, $(4, 3)$ and $(\frac{1}{4}, \frac{1}{2})$, we find that

$$4^2 + 3^2 > 1$$

$$\left(\frac{1}{4}\right)^2 + \left(\frac{1}{2}\right)^2 < 1$$

where the symbols $>$, $<$ are read as "is greater than" and "is less than," respectively. It is clear that $(4, 3)$ and $(\frac{1}{4}, \frac{1}{2})$ are not among the points on the circumference of our circle. If, in fact, we locate them on the plane pictured on page 69, we can actually see that $(4, 3)$ would fall outside of a circle of radius 1 about the origin while $(\frac{1}{4}, \frac{1}{2})$ would fall inside. Thus, with the equation for the circle already given and various related equalities, we are able to divide the points on the plane into various sets:

A. $x^2 + y^2 = 1$ the set of points on the circumference

B. $x^2 + y^2 < 1$ the set of points interior to the circle

C. $x^2 + y^2 > 1$ the set of points exterior to the circle

D. $x^2 + y^2 \leq 1$ the set of points on the circumference and the interior of the circle

E. $x^2 + y^2 \geq 1$ the set of points on the circumference and the exterior of the circle

F. $x^2 + y^2 \neq 1$ the set of all points not on the circumference

Certain pairs of these sets, when combined, will include all of the points in the plane and yet will have no points in common: (A) and (F), (B) and (E), (C) and (D). These are called complementary sets.

When we divide the entire plane into such parts, even though we cannot physically represent some of them, like the points on the circumference or the interior of the circle without the circumference, we are still dealing with the concept of the whole and its parts in the traditional manner. The plane is the sum of its sub-sets (A), (B) and (C); each occupies a "different" portion of the plane. Yet with point sets it is possible to divide the plane into various pairs of complementary sets in such a way that each set of the pair by itself is everywhere dense upon the plane. Such a pair would be the set of all points in the plane which have rational coordinates; and its complement, the set of all points which have at least one irrational coordinate. Together, they include all the points in the plane, which are everywhere dense. Yet, when we remove either set of points, the points remaining are still everywhere dense in the plane. This curious situation arises from the fact that the rational numbers are everywhere dense (i.e., between any two rational numbers there is always another rational number) and that the same characteristic is exhibited by the irrational numbers.

There is yet another unconventional way in which we can divide the whole point set into parts, or sub-sets of points, a way which is not available to us when we are dealing with geometrical figures in the traditional manner. As we have seen, we can divide a point set into a finite number of complementary sets, or parts; but we can also divide it into an infinity of such parts. The number of points on a line, in a plane or in a space is always the same: a non-denumerable infinity. If we divide any one of these point sets into sub-sets, each of which contains but a single point, we have divided the whole into a non-denumerable infinity of parts.

Such a non-denumerable infinity is infinitely more numerous than a denumerable infinity; yet we can also divide a point set which contains a non-denumerable infinity of points into a denumerable infinity of sub-sets. Later we shall see that this is sometimes a rather complicated procedure, but now we shall merely divide the real-number line into a denumerable infinity of parts. This is child's play in the theory of point sets. By defining each sub-set as all the real numbers equal to and greater than a given integer n but less than the next largest integer, or $n + 1$, we have solved the problem. The integers, a denumerable infinity themselves, divide the non-denumerable infinity of real numbers, which represent all the points on the line, into a denumerable infinity of sub-sets, each of which of course contains in turn a non-denumerable infinity of points.*

* The reader is reminded of the proof on page 218 and the following pages that the real numbers between 0 and 1 are a non-denumerable infinity, and of the proof on page 221 that the number of points on any portion of the line is equal to the number of points on the entire length of the line.

The distinction between non-denumerable and denumerable infinities, as confusing as it may be to us at first, is essential to our gaining even a glimpse of the reasoning which leads to the paradoxes of point-set theory and their implications for the theory of measure. We must, therefore, make sure that we have it clearly in mind before we go any further in this chapter. We recall from Chapter 14 that a denumerable or countable infinity (the "smallest" of all infinities) is one whose members can be placed in one-to-one correspondence with the integers, and thus —in the sense that there is an ordered pairing between its members and the integers—can be counted. Such countable infinities include the integers themselves; such subsets as the natural numbers, the even numbers, the primes, and so on; and, what is particularly important to us in point-set theory, the rational numbers. A non-denumerable infinity, as we saw in the same chapter, is more numerous than the integers, cannot be arranged in any way so that its members can be paired with them, and hence cannot be "counted" in the same sense that a denumerable infinity can be counted. Such uncountable or non-denumerable infinities include the real numbers—which singly, in pairs, or in triples can be placed in one-to-one correspondence with the points of line, plane and space, respectively. They also include a non-denumerable sub-set of the reals which is particularly important for point-set theory—the irrational numbers. It is essential that we keep in mind the fact that while the rationals and the irrationals are complementary sub-sets of the real numbers, the rationals are denumerable and the irrationals are non-denumerable.

In brief summary:

1. Each of the geometrical figures, plane and solid, with which we shall deal in the next few pages contains a non-denumerable infinity of points.

2. Each and every one of such a non-denumerable infinity of points can be handled as an individual because it can be uniquely defined by ordered real-number coordinates.

3. The real numbers, which are the rational numbers plus the irrational numbers, are a non-denumerable infinity.

4. The rational real numbers are a denumerable infinity.

5. The irrational real numbers are a non-denumerable infinity.

We are now prepared to follow the reasoning which will lead us to a fundamental paradox of point-set theory:

The whole is not necessarily greater than one of its proper parts, but on the contrary can be congruent to that part.

The word *congruent* here means "equal" in that special sense in which we use it in the geometry with which we are all familiar. In point sets we always use it in this sense. As a specific example, we say that the triangles A and B below are congruent if, without lifting the left-hand triangle out of the plane, we can, by rigid motion alone (sliding along the page in this case), superpose it upon the right-hand triangle so that the two occupy exactly the same position and there is a one-to-one correspondence between their points. The triangle C, as can be seen, is a proper part of A; but since A can never, by rigid motion alone, be superposed on C, they are not congruent.

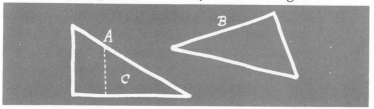

In point-set theory the meaning of the word *congruent* is exactly the same as it is in traditional geometry—superposition and one-to-one correspondence achieved by rigid motion alone. But here the resemblance stops. For in traditional geometry we never find, as we do in point sets, that the whole can be congruent to its proper part. We can never superpose *A* in the figure above upon *C*, its proper part; but we can superpose the whole right-hand half of the plane, or the set of all points such that *x* > 0, upon a proper part, the set of all points with *x* > 1:

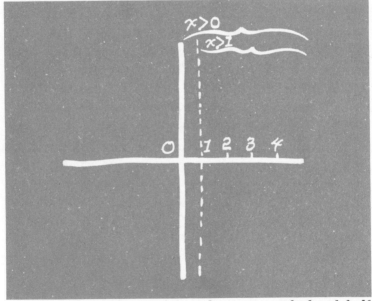

It is "obvious" to us that the entire right-hand half of the plane (*x* > 0) is "larger" than that "part" of it (*x* > 1) which lies to the right of 1, "larger" in the same way that triangle *A* is larger than triangle *C*. Yet, recalling Cantor's theory of the infinite, we know that it is perfectly possible for an infinite set (such as the integers) to be equal (because placed in one-to-one correspondence with

234

it) to a proper part (such as the even numbers). It is only a step to the recognition that the half-plane of points can be superposed on its proper part because the points of each can be placed in one-to-one correspondence merely by sliding the whole onto its part. Since such superposition achieved by rigid motion is the accepted definition of congruence, we can say in this situation that the whole is congruent to its proper part.

In point-set theory this same notion of congruence is found in sets much more complicated than the points of the half-plane. For an example of such a set, we begin by marking off on a circle an angle which is an irrational multiple of one complete rotation of the circle, or 360°. If we were to make our angle a rational multiple (for instance, 90° or one-fourth of a complete rotation), we would find that after we had marked off four angles our next would coincide exactly with one which we had previously marked off. When, however, our angle is irrational, like

$$\frac{1°}{\sqrt{2}}$$

no matter how many times we go around the circle we shall never mark off an angle which coincides exactly with one which we have previously marked off.

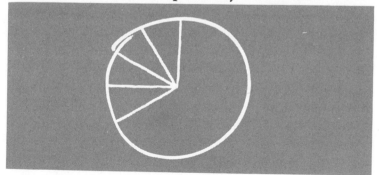

We thus divide the circle into an infinite number of line segments—in this case a denumerable infinity, since each segment can be paired in order with an integer, the first with 1, the second with 2, and so on. The point set which we extract in this way from all the points of the circle consists of a denumerable infinity of line segments, each of which contains a non-denumerable infinity of points. Now this point set can be shown to be congruent to a proper part. By rigid motion—in this case a rotation of the whole point set through the distance of our chosen angle—we bring each segment into the position originally occupied by the next segment in the construction. Since, however, no segment can have been brought into the position originally occupied by the first segment, we have shown that the whole is congruent to its proper part.

The same result could have been achieved by considering as our point set the end points of the segments which lie on the circumference of the circle. Then by rotating the circumference we would place the first point on the second, the second on the third, and so on. A numerical equivalent would be the placing of the positive integers in one-to-one correspondence with the positive integers greater than 1:

1	2	3	4	5	6	7	...
↓	↓	↓	↓	↓	↓	↓	
2	3	4	5	6	7	8	...

For our next example, instead of extracting a denumerable infinity of points from the circumference, we shall divide the entire circumference into a denumerable infinity of congruent pieces. This is not, by any manner of means, child's play. The difficulty lies in the phrase "a denumerable infinity." It is no problem at all to divide the circumference of any circle into a finite number of congruent pieces. We can, for instance, use the length of the radius to mark off six arcs on the circumference, any one of which can be superposed by the rigid motion of rotation on any one of the others.* Nor is it a problem, as we have already seen in connection with the real-number line, to divide the circumference into a non-denumerable infinity of congruent pieces, since when each piece consists of just one of the non-denumerable infinity of points on the circumference, all of them together are a non-denumerable infinity of congruent pieces. Here, however, the method we used to divide the real-number line into a denumerable infinity of pieces, which were integer intervals, will not work, since the real-number line is infinite in length while the circumference of the circle is finite. To solve our problem we must resort to a much deeper type of reasoning.

We begin with a circle the circumference of which is one unit in length. The circumference can then be thought of as the portion of the real-number line between 0 and 1. Except for the fact that 0 and 1 are the same point, all the other points on the circumference are uniquely identifiable as the real numbers from 0 to 1.

To divide this *non-denumerable* infinity of points on the circumference into a *denumerable* infinity of sub-sets, we first gather together a set (or what we shall call a family, to distinguish it from the other types of sets) which consists of all those points that differ from some point on the circumference by a rational number, or distance. The first family consists of all those points which are a rational

---

* Even in such a simple problem as this, in point sets we have to decide how to distribute the end points of the arcs, since each shares its end points with the adjacent arcs. (We usually go around the circle counterclockwise and assign to each arc its first end point.)

distance from the point 0. This family, we can see, will include all the rational points on the circumference since $0 + \frac{1}{2}$ gives us the rational point $\frac{1}{2}$; $0 + \frac{1}{3}$ gives us the rational point $\frac{1}{3}$, and so on. We do not need to bother with selecting points a rational distance from any rational point other than 0, for the points so selected would necessarily be duplicates of those already included in the first family (the sum of rational numbers being always a rational number). We turn our attention to selecting families of points which are a rational distance from each irrational point in turn. One of these families, for instance, will be all those points a rational distance from the irrational point $1/\sqrt{2}$. Since we are choosing a different family of points for each irrational point and since there is a non-denumerable infinity of such irrational points between 0 and 1, we shall divide the non-denumerable infinity of points on the circumference into a non-denumerable infinity of families, or point sets. How many points in each of these families? Only a denumerable infinity, for there is just one point for each rational distance and the rationals themselves are a denumerable infinity.

From the families, we now gather together a new kind of point set which we can call, to distinguish it from a family, a set of representatives from each family. The first set of such representatives is obtained by choosing from each of the non-denumerable infinity of families a single point, and will thus contain a non-denumerable infinity of points.* The next set of such representatives is obtained

* After all that we have accepted so far, we probably have no difficulty in accepting the idea that we can choose from each of a non-denumerable infinity a single point. Yet this statement—known as the Axiom of Choice—has been one of the most controversial in modern mathematics. It is easy to see that if we have a finite number of sets, no two of which have a common member, we can in a finite number of operations choose a member from

by rotating the entire circle a given rational distance and taking from each family a second point which is that rational distance from the first. To obtain yet another set of representatives we again rotate the circle a different rational distance; we continue in this manner until we have a set of representatives for each of the denumerable infinity of rational distances on the circumference.

It is logically clear (although it may take a moment for one unused to juggling infinities to see that it is) that we will end up with a denumerable infinity of sets of representatives, for there will be one for each rational distance—a denumerable infinity. Each set of representatives, however, will contain a non-denumerable infinity of points, one from each of the non-denumerable infinities of families we first selected.

None of the sets of representatives can have a point in common with any other set because each rotation gave us a choice which, by its nature, could not include any of the points selected by previous rotations. Since the points

each set so that we have a new set which has just one member in common with each of the original sets. If, however, we have an infinite number of sets to choose from, we cannot choose the new set in a finite number of operations unless we have some way of automatically distinguishing the member to be chosen. This difficulty is illustrated in Bertrand Russell's story of the infinitely rich man with infinitely many pairs of shoes and socks. He can easily form a set which has one member in common with each pair of shoes. The rule for membership in this set can be that each member must be a *left* shoe. With the statement of this rule, the set is automatically chosen. In the case of the socks, however, no such rule is possible. A sock *must actually be chosen* from each pair since socks, unlike shoes, are not automatically distinguishable as "left" or "right." Since even our infinitely rich man could never complete this infinite task, the set containing one sock from each pair could never be chosen. Mathematicians usually overcome this difficulty with the Axiom of Choice by means of which they simply *assume*, as an axiom, that it is always possible to choose one member from each of an infinite number of sets.

which we are choosing constitute a denumerable infinity, as does each of the families from which they are being chosen, every point on the circumference will be included in some set of representatives. We have, therefore, by dividing the circumference into mutually exclusive sets including every point, divided the circumference into a denumerable infinity of pieces. These pieces are congruent in the sense of elementary geometry, for all were obtained by the rigid motion of rotation. We have solved the given problem: *to divide the circumference of a circle into a denumerable infinity of congruent pieces.*

The significance of what we have done may not be immediately apparent to the reader whose head is still rocking with non-denumerable and denumerable infinities; but let us consider for a moment the problem of assigning a measure, or a length, to these pieces of the circumference. Among them are included all the points on the circumference. By everyday standards they are the parts of the circumference and the circumference is the whole, so the sum of their lengths should be the length of the circumference. But by everyday standards they are also congruent, or equal. If, in an everyday sense, any measure is assigned to the pieces, the same measure must be assigned to each one of them. There are two possibilities: either a measure of 0 for each piece or a positive measure. The circumference of the circle is one unit, and the pieces into which we have divided it must, if they are to have any length, add up to 1. Yet the sums of the only measures we can possibly assign to them are zero or infinity. We are forced to the necessary conclusion that these pieces— the congruent point sets into which we have divided the entire circumference—*do not have a length.*

The problem which we have just detailed rather completely is an example of the type of reasoning, although much less deep, which led to the most famous paradox of point-set theory and an implication in regard to everyday ideas of measure much more startling than the one above. The Banach-Tarski paradox was propounded in 1924 by Stefan Banach (1892-1945) and Alfred Tarski (1901-   ). These two mathematicians proved that it is possible to disassemble a solid unit sphere into a finite number of pieces in such a way that the pieces could be reassembled into two spheres the same size as the original sphere!

Mathematically, the most unusual thing about the Banach-Tarski work was that its paradox of measure rested, not upon an infinity of pieces, as in the case of the problem we have just finished examining, but on a down-to-earth *finite* number of pieces. How many pieces? They did not say. A very large number of pieces? They did not say. Merely a finite number of pieces. That in itself was sufficiently startling.

The exact number of pieces necessary was given, some twenty years later, by R. M. Robinson (1911-   ), and it was very small. Working with only five pieces, Robinson showed it is possible to disassemble a solid unit sphere (point by point, of course) and reassemble it into two spheres the same size as the original. The reasoning which led Robinson to this conclusion was very complex, but basically similar to that which we followed in dividing the circumference of a circle into a denumerable infinity of congruent pieces to which no length could be assigned.

In determining the smallest finite number of pieces into which the solid sphere can be divided for the Banach-Tarski paradox, Robinson began with the simpler problem of determining the number of pieces into which the surface of such a sphere—or a hollow sphere—must be divided so that it could be reassembled into two spheres the same size as the original. He showed how it was possible to divide the point set of the surface into four sub-sets A, B, C and D which exhibit a truly remarkable property. The sub-sets A and B are congruent to each other; and each of them is also congruent to the sum of A and B. In the same way C and D are congruent to each other and each of them, to the sum of C and D. Thus by rotating A into $A + B$ and C into $C + D$, we are able to form S_1, a sphere which is exactly like our original sphere. We then rotate B into $A + B$ and D into $C + D$ to form S_2, a second sphere exactly like S_1 and hence exactly like our original sphere. Thus four pieces were shown to be sufficient for reassembling a hollow sphere into two spheres the same size as the original.

The solution of the problem for the solid sphere was then shown by Robinson to be essentially the same as that for the hollow sphere. Yet there was a difficulty. We can of course extend the four pieces of the surface A, B, C and D into the center of the sphere, but which piece will then include the point which is the center? If we are willing to simply assign the point to one of the four pieces so that it has one more point than the others, then we can reassemble A, B, C and D into two solid spheres exactly like the original except for the fact that one of the new spheres will not have a point at its center. Most of us would be satisfied with this solution, but a mathematician will go to considerable trouble to get a center for that other sphere. Having found a point by a method too devious to record here, Robinson brought it to the center of the sphere by translation * (all the other rigid motions in-

* *Translation* is distinguished from *rotation* in that, under the rigid motion of translation, all the points are moving in the same direction at the same time.

volved in the solution being rotations about the origin which of course could not produce the needed copy of the origin). Five, then, was determined as the necessary and sufficient number of pieces for the Banach-Tarski paradox.

The significance of this paradox for the theory of measure is immediately apparent. When we consider geometric figures as point sets in 3-dimensional space and we do nothing more to them than what we do to the usual run of geometric figures with which we are familiar, we are forced to the conclusion that we cannot assign to them a measure of either area or volume. If the four pieces into which the surface of our sphere was divided had an area, their sum would be both the area of the original sphere and twice the area. If the five pieces of the solid sphere each had a volume, their sum would be both the volume of the sphere and twice the volume. In these particular situations the sum of the parts is not the whole, but twice the whole!

A conclusion like this—completely contrary to everything our intuition tells us, to what we have always known with confidence that we knew, and to what we feel is true—separates the mathematical minds from the inherently non-mathematical. For there are always those who want to go back to the beginning, change the rules, forbid such exceptions, refuse such conclusions. The man who was the founder of point-set theory was not one of these.

Georg Cantor came to the theory of point sets because he was *forced*—this was his own word for it—by logic. He did not invent his theory, arbitrarily, to confound intuition and experience. It is indeed one of the neatest ironies of mathematics that this theory, which seems as completely removed from the practical world as do the dreamy speculations of Laputan philosophers, grew out of the work of Jean Baptiste Joseph Fourier (1768-1830), a physicist who expressed his opinion frequently and positively that mathematics justified itself only by the help it gave to the solution of physical problems. (Fourier's own considerable contributions to mathematics were in the theory of functions, and resulted from his researches in the conduction of heat.) Although the line from Fourier to Cantor is a direct one, it is not the whole line. The theory of point sets is more truly a modern step on a logical path to which mathematics committed itself when it accepted the idea that there is a measure for every length—a real number, rational or irrational, for every point on the number line.

Georg Cantor followed this path where it logically led and drew the necessary conclusions although they were contrary to his own intuition, training and desire, and made him the object of an attack which had been unequaled, in mathematics, since the Pythagorean who discovered the irrationality of $\sqrt{2}$ perished, mysteriously, at sea.

16

*The
New Euclid*

FOR OVER TWO THOUSAND YEARS THE *Elements* of Euclid commanded the almost unqualified admiration of mankind. It could be said—and was:

". . . from its completeness, uniformity and faultlessness, from its arrangement and progressive character, and from the universal adoption of the completest and best line of argument, Euclid's *Elements* stands preeminently at the head of all human productions."

It could be added—and was:

"For upward of two thousand years it has commanded the admiration of mankind, and that period has suggested little toward its improvement."

At the beginning of the twentieth century, however, the suggestion box was open.

This was the period known in the history of mathematics as "the crisis in foundations." A quarter of a century had elapsed since Georg Cantor had presented his theory of the infinite; and mathematics, somewhat like a man with a new living-room chair, had at last settled back comfortably with the once revolutionary idea of the infinite as something consummated. This was the moment that the Italian mathematician C. Burali-Forti (1861-1931) chose to produce—by using exactly the type of reasoning that Cantor had used to establish his theory of infinite sets—

245

a flagrant contradiction which, at least for the moment, virtually invalidated Cantor's entire theory. The new chair collapsed; and, of course, like any normal man, mathematics in general now refused to use any chair but that chair.

The effort to set logically aright the foundations of mathematics and yet retain Cantor's new theory of the infinite ("No one shall expel us from this paradise which Cantor has created for us!") was led by a great German mathematician who now occupied Gauss's old place at Göttingen. David Hilbert (1862-1943) was actually the greatest mathematician in the world during the time that the newspapers and the man on the street thought unquestioningly it was Einstein (who was not a mathematician but a physicist). Besides the notable work which Hilbert accomplished in several fields, he offered mathematics leadership at a time when it was desperately needed.

Faced with the crisis in foundations, Hilbert led his followers back to the Greeks, back to Euclid, to begin in an almost literal sense at the foundations themselves and re-erect the edifice of mathematics, block by block, with modern rigor.

While the *Elements* of Euclid had served as the model of logical thought since antiquity, it had been observed by various mathematicians during that time that there were, nevertheless, certain logical lapses in the logical model. In the very first proposition of the very first book one such flaw is immediately apparent to the rigorous eye. Euclid lays the first block of the edifice of elementary geometry by attempting to show that (relying only on the previously stated definitions and axioms) it is possible "on a given finite straight line to construct an equilateral triangle." In his proof, invoking Postulate 3, he inscribes a circle with center A and radius AB on the given segment

AB and another circle of the same radius with center *B*. He then proceeds with his proof from the point *C* "in which the circles cut one another."

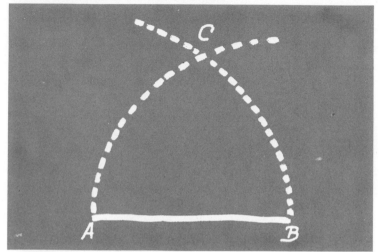

"It is a commonplace," says Sir Thomas Heath, rather tiredly, in his English edition of the *Elements*, "that Euclid has no right to assume, *without premising some postulate*, that the two circles *will meet* in the point *C*."

Like Euclid, we think that we know—for haven't we drawn this same figure many times to find the center of a given line?—that the circumferences will intersect at a point equidistant from *A* and *B*. We cannot *know* this from experience, for we cannot have drawn all possible circles; we can only *assume* from what experience we have had that such pairs of circles will always meet and that they will always meet in just one point above the line. This, then, is an assumption upon which our geometry is based and, as such, it should be stated with the other assumptions. Because such assumptions of intersection are not explicitly stated in the *Elements*, it is possible by using only Euclid's stated definitions and axioms to "prove" such paradoxical propositions as "Every triangle is an isosceles triangle." (The "proof" is given in J. W. Young's excellent little book *Fundamental Concepts of Algebra and Geometry,* Macmillan, 1930.)

Unstated assumptions of intersection pervade the *Elements;* for Euclid uses whenever possible the method of actual construction for proving the existence of figures having certain properties. Constructions are made on the basis of Postulates 1-3, with straight line and circle alone. What enables Euclid to build with these straight lines and circles is the fact that they determine *by their intersection* other points in addition to those originally given in the problem, and that these points can then be used to determine new lines, and so on. The method of the *Elements* logically demands that the existence of such points of intersection be either proved or postulated in the same way that the existence of the lines which produce them is postulated by 1-3.* The ladder to truth rests on the idea that *after the original assumptions are granted, no other assumptions will ever be required.*

Hilbert's problem in subjecting the *Elements* to true modern rigor was, not only that Euclid often assumed assumptions which he had not made, but that he relied upon definitions which did not actually define. We are all familiar with this problem of definition. We define an orange—with Webster—as "the nearly globose fruit, botanically a berry, of an evergreen rutaceous tree (genus *Citrus*)"; and immediately find ourselves involved in a multiplying set of definitions: What is a fruit? What is a globose fruit? What is a nearly globose fruit? It is obvious that unless we can begin with the assumption that there are certain terms which everybody intuitively "knows," we shall have to give up our project of a dictionary.

---

* The only statement of intersection in the axioms is the negative one of the Parallel Postulate, where it is stated that under certain conditions two lines will never meet.

Euclid faced the same problem as the dictionary maker when he went to compile the *Elements*. He began in traditional style with first things first: twenty-three definitions at the beginning of Book I, ranging from "A point is that which has no part" to a breakdown of such quadrilateral figures as the square, the oblong, the rhombus and the rhomboid. "And let quadrilaterals other than these be called trapezia," he concluded. Although Euclid grouped these definitions together, he made in the *Elements* a distinction, never stated but clearly implied. The existence of points, lines and circles had to be assumed by the reader; existence could not be proved for any of these. But after the existence of these geometrical objects was assumed, such a figure as met the requirements of a rhomboid, for instance, could be constructed and displayed to the reader, its "existence" established by proof.

Although a rhomboid is perhaps a vague figure to the ordinary person,* points, straight lines and circles are not. Everybody who has seen a point on paper, for instance, knows intuitively what Euclid meant by his geometric point "that has no part."

The ancients argued quite extensively about the proper definition of a point:

A point is an extremity of a line.
A point is that which is indivisible.
A point is that with position only.
A point is an extremity which has no dimension.
A point is the indivisible beginning of all magnitudes.

Yet all of their definitions were attempts to express what they thought they already "knew" a point was.

* What is it? A rhomboid, according to Euclid, is "that which has its opposite sides and angles equal to one another but is neither equilateral nor right-angled."

Actually the definitions were not really necessary at all. Anyone who had never seen a point on paper would, by the time he had completed the propositions of the *Elements,* have a thoroughly accurate idea of the geometric point as a result of the various statements which are made about it in the propositions.

From what we have said about the problem of definition, three main ideas emerge—as pertinent for modern mathematicians as for Euclid:

1. Unless we are to continue defining indefinitely, some terms in our geometry will have to be accepted as undefined, or *primitive,* terms.

2. Their existence will have to be assumed—just as the statements made in the axioms have to be assumed because they cannot be proved without bringing in other axioms which cannot be proved without bringing in other axioms, and so on.

3. What these undefined or primitive terms stand for will, however, become increasingly clear as—using only these terms, the axioms, and the rules of reasoning—we make and prove more and more statements about them.

Now let us return to Hilbert and his effort to place at least one domain of mathematics—the geometry of Euclid—on a thoroughly sound logical basis. (This effort, massive though it was, was presented in a tiny book which is available in English as *The Foundations of Geometry* [Open Court, La Salle, Illinois, 1938].)

As the epigraph of his work, Hilbert took a quotation from the great German philosopher Immanuel Kant (1724-1804). It was Kant whose often quoted attitude toward the

250

axioms of the original Euclid was that they were "a priori synthetic judgments imposed upon the mind, without which no consistent or accurate reasoning would be possible." Since the time when Kant made this statement, it had become increasingly clear—in mathematics, if not in philosophy—that Euclid's assumptions were somewhat arbitrary and that other assumptions and other geometries were just as possible and just as "true." The quotation from Kant which Hilbert chose for his epigraph was not, therefore, this most famous one. He chose instead a statement to emphasize the relation between the intuitive roots of mathematics and its abstract flowering:

"All human knowledge begins with intuitions, thence passes to concepts and ends with ideas."

The intuitions with which geometry begins, in both a literal and a figurative sense, are those of points and lines and the surfaces on which points and lines exist. These, then, are the ideas with which both Euclid and David Hilbert begin. But every one of the twenty-three hundred years which lie between the Greek and the German lie between their opening treatments of these ideas. To emphasize the contrast, we shall present their two beginnings in parallel columns:

EUCLID	HILBERT
Definitions	*The Elements of Geometry*
	Let us consider three distinct systems of things.
A *point* is that which has no part.	The things composing the first system, we will call *points* and designate them by the letters $A, B, C, \ldots$;

251

A *line* is breadthless length. The extremities of a line are points. A *straight line* lies evenly with the points on itself.

A *surface* is that which has length and breadth only. The extremities of a surface are lines. A *plane surface* is a surface which lies evenly with all the straight lines on itself.

those of the second, we will call *straight lines* and designate them by the letters $a, b, c, \ldots$;

and those of the third system, we will call *planes* and designate them by the Greek letters $a, \beta, \gamma, \ldots$.

The points are called the *elements of linear geometry;* the points and straight lines, *the elements of plane geometry;* and the points, lines and planes, *the elements of geometry of space* or *the elements of space.*

Euclid continues to a total of twenty-three definitions at the beginning of Book I. Hilbert is through with the definition of terms.

But where the new Euclid is more concise in his definitions than the old, he finds a need for many more assumptions than the five Common Notions and five Postulates of the original *Elements.* In Hilbert's Euclid there are twice as many axioms, and the relationship between the axioms and the three undefined "systems" listed above is explicitly stated:

252

We think of these points, straight lines and planes as having certain mutual relations, which we indicate by means of such words as "are situated," "between," "parallel," "congruent," "continuous," etc. The complete and exact description of these relations follows as a consequence of the *axioms of geometry*. Each of these groups expresses, by itself, certain related fundamental facts of our intuition. We will name these groups as follows:

> I. 1–7. Axioms of *connection*.
> II. 1–5. Axioms of *order*.
> III. Axiom of *parallels*.
> IV. 1–6. Axioms of *congruence*.
> V. Axiom of *continuity*.

The relationship between the three "elements" of his geometry and the twenty "axioms" is precisely stated by Hilbert within the paragraph quoted above: "The complete and exact description of these relations [between the elements] follows *as a consequence* [my italics] of the *axioms of geometry*."

For example, in the first group of axioms which have to do with the intuitive idea of "connection," he begins with the assumption that "two distinct points *A* and *B* always completely determine a straight line *a*." He then points out that instead of the word "determine," we may also employ other forms of expression: "For example, we may say *A* 'lies upon' *a*, *A* 'is a point of' *a*, *a* 'goes through' *A* 'and through' *B*, *a* 'joins' *A* 'and' or 'with' *B*, etc."

Within the body of the work, Hilbert places various necessary "definitions" as the need for them occurs. These are made in terms of the previously stated but undefined elements, point, straight line and plane, and the description of their relations which follow as a consequence of

253

the axioms. Such a definition is that of *segment:*

DEFINITION. We will call the system of two points *A* and *B,* lying upon a straight line, a *segment* and denote it by *AB* or *BA.*

In the preceding paragraphs we have given a few quotations directly from Hilbert's work on the foundations of geometry so that the reader can feel at first hand at least the faint breath of the new spirit of rigor which entered mathematics at the beginning of the twentieth century. It was by such rigor—no attempt made to define the undefinable, but every attempt made to state explicitly every necessary assumption—that Hilbert attempted to resolve "the crisis" by setting the most ancient branch of mathematics upon a logically sound foundation.

The crisis had been precipitated by the discovery that the type of reasoning and the assumptions Georg Cantor had used in developing the theory of sets could lead to *contradiction.* To resolve the crisis, Hilbert had returned to Euclid and attempted to place elementary geometry on a completely rigorous foundation which would eliminate such contradictions as the paradoxical proposition that all triangles are isosceles. He then attempted to do something which would never have occurred to Euclid. He set out to *prove* that reasoning with his assumptions could not possibly lead to such contradictions—that the axioms of elementary geometry as he had restated them were now absolutely consistent!

How is it possible to prove that a set of assumptions is consistent? How can we know before we start that we will never find ourselves in the position of having proved that *A* is equal to *B* and, also, that *A* is not equal to *B?*

At the present time the only way of doing this is to

254

match our abstract theory—primitive terms and assumptions—with some concrete representation of it, which is already granted to be consistent. For instance, we can take as our model of consistency the arithmetic of real numbers, since reasoning according to the rules has never yet brought us to the contradictory position of having proved that $2 + 2 = 5$ as well as 4!

This is what Hilbert did, although taking an even smaller domain of arithmetic than that of the real numbers. This was the domain X "consisting of all those algebraic numbers which may be obtained by beginning with the number one and applying to it a finite number of times the four arithmetic operations (addition, subtraction, multiplication and division) and the operation $\sqrt{1 + w^2}$, where w represents a number arising from the five operations already given." The reader will recognize this domain as that of the constructible numbers, which we met in Chapter 9.

In the terms of the arithmetic of this domain, Hilbert defined his primitive terms, point, straight line and plane. A pair of numbers (x, y), for instance, became a point and the ratio of three such numbers $(u:v:w)$, where u and v are not both equal to 0, became a straight line. The existence of the equation

$$ux + vy + w = 0$$

was defined to express the condition that the point (x, y) lies on the straight line $(u:v:w)$. He then showed how the various groups of axioms could be interpreted in the terms of the arithmetic of domain X. In this way he was able to establish that the arithmetic of domain X could be considered a concrete representation of his abstract geometry of three "systems" and twenty "axioms."

"From these considerations," he concluded, "it follows that every contradiction from our system of axioms must also appear in the arithmetic related to the domain X."

We may still ask, "But how did Hilbert know that the arithmetic of real numbers is consistent?" The answer is that he did not know. No one knows. The arithmetic of real numbers is considered to be consistent only because of the absence of any known contradiction. That it is in actuality consistent was an assumption that Hilbert made. He was aware that it was an assumption. Since the consistency of his geometry depended upon the consistency of the arithmetic of real numbers, only an absolute proof that the arithmetic is consistent would establish the absolute consistency of his geometry. Although there had not yet been such an absolute proof when Hilbert published his work, one was generally assumed to be possible.

This was as far as David Hilbert could go in his effort to resolve the crisis in foundations by establishing the logical consistency of elementary geometry. At the time he wrote the words above, he was almost forty, the twentieth century was in its first year, and not yet born was the young man who would reveal the hopelessness of Hilbert's dream by demonstrating with finality that establishing the absolute consistency of any such set of axioms is *impossible*.

In 1931, at the age of twenty-five, Kurt Gödel published a paper entitled "On Formally Undecidable Propositions of *Principia Mathematica* and Other Related Systems." When, many years later, Harvard University awarded him an honorary degree for this work, the citation referred to him as "discoverer of the most significant truth of this century, incomprehensible to laymen, revolutionary for philosophers and logicians."

For the moment we shall not be concerned with these

256

even wider implications of Gödel's 1931 paper, but only with its primary subject. This was the demolishing of the hope that the absolute consistency of any mathematical system (including ordinary arithmetic) could be established.

Although Gödel's is as complex a piece of reasoning as mathematics is ever likely to see, it depends upon a variation of an ancient brain teaser with which we are all familiar. This is the statement of Epimenides, who was a Cretan, that all Cretans are liars. Was Epimenides a liar?

In his epochal proof Gödel showed that in any sufficiently strong mathematical system it is possible to construct a statement which asserts its own unprovability in that system. The consistency of the system cannot then be established within the system itself but must be referred to a stronger system, where of course the same thing can be shown to be true, so that the consistency of that system must be referred to a still stronger system, and so on.

With the kind of finality which is possible only in mathematics, Gödel demolished Hilbert's project. There can be no proof of the absolute consistency of the foundations of mathematics. We must live and work on *assumptions* of consistency.

David Hilbert died in 1943 at the age of eighty-one. The problem which he had put for himself was one that would never have occurred to Euclid. The answer which young Kurt Gödel established in his epochal paper of 1931 was one that had never occurred to Hilbert.

257

17

Of
Truth Tables
and Truths

THE LADDER TO TRUTH WHICH EUCLID erected in the *Elements* consisted of the rungs of definitions, axioms, theorems and proofs suspended between strong side supports of logic. These supports were formulated in Euclid's day as the laws of reason—and they were formulated *in words,* for they were not part of mathematics but of logic. Today these laws are still the strong and indispensable supports of the ladder to truth, but today they are expressed in mathematical symbols; and any proof which utilizes a combination of these laws can be tested for error by a mathematical method—the method of the truth tables.

The truth tables are a development of the sentential, or propositional, calculus. The sentential calculus, in spite of its formidable name, has a vocabulary which consists in its entirety of the small words, *and, or, not, if, then, only,* and the one relatively big word, *sentence.* It is a fragment—and we must admit the most elementary fragment—of a great and modern mathematical study—symbolic logic—which subjects logic to the symbols and procedures of mathematics.

The basic logical concepts of the sentential calculus are things which every mathematically minded person knows and uses intuitively. They sound, therefore, too obvious to bother with. But as mathematical sentences (or propositions) become longer and more complicated, intuition is not sufficient to determine with finality their logical truth or falsity. Then a method is needed which is completely formal, and this method is furnished to mathematics by the sentential calculus under the slightly sinister title of *truth tables*.

We shall, in the course of our exploration of the sentential calculus, use the method of truth tables to test the logical truth of certain statements; but before we can do so we must examine in some detail the meaning of its vocabulary and familiarize ourselves with the five symbols with which it conducts its business. The reader is strongly urged to do the simple problems as they occur, covering the answers with his hand and testing his memory of what has been explained, translating language into logical symbolism and logical symbolism back into language, and taking pencil in hand and determining for himself the truth of given sentences. It is guaranteed that he will be pleasantly surprised at the enjoyment he will get out of actually using truth tables.

The vocabulary of the sentential calculus is, as we have said, limited to very simple and common words. These words are used, however, in a precise way which, in every case, seems different either to a large or small degree from the way in which we ordinarily use them. Because we use the words of the sentential calculus all the time, we have a tendency to feel that, like Humpty Dumpty, we have as much right as anybody to say what they mean. We are inclined to object to the meanings which the logicians assign to them. (Even logicians have this same feeling about the words and argue quite a bit among themselves.) But if we are to understand, we must make a definite effort to erase from our minds our own

259

personal meanings of the words which compose the vocabulary. We must consider these words as technical terms to which the logician, like any scientist, assigns the unambiguous definitions which are necessary for the functioning of his science.

The most straightforward way of getting rid of the ordinary meanings of the words is to eliminate the words themselves from our preliminary discussion. So let us begin by giving our attention to the five symbols of the sentential calculus, each of which represents a logical concept.

$$\sim \quad \wedge \quad \vee \quad \rightarrow \quad \longleftrightarrow$$

$\sim$ (*Negation*). This symbol, when placed before a sentence, or a letter which we take to represent a sentence, denies whatever follows it. If we represent a sentence by a variable p, then $\sim p$ stands for "not p." If the sentence p is "Snow is black," then $\sim p$ is "Snow is not black." We can *call* this logical concept "not."

$\wedge$ (*Conjunction*). This second symbol indicates the joining together of the expressions on either side of it. If these expressions are variables p and q, then $p \wedge q$ indicates "p and q." If p is used as above and q in place of another sentence, such as "All men are mortal," then $p \wedge q$ is the sentence—what we call grammatically a compound sentence—"Snow is black, and all men are mortal." We can call this logical concept "and."

$\vee$ (*Disjunction*). The third symbol represents a joining which nevertheless leaves the joined expressions somewhat separated. This is sometimes called an alternation. If $\vee$ is used to join our two variables p and q, the resulting expression $p \vee q$ is the equivalent of "p or q." In the case of the meanings we have been assigning to the variables, the expression can be translated as "Snow is black, or all men are mortal." This is called the logical concept "or."

260

→ (*Implication*). With this fourth symbol we have what is grammatically called a conditional sentence. The expression $p \rightarrow q$ is read "If p, then q," or, "If snow is black, then all men are mortal." This logical concept is called "if, then."

⟷ (*Equivalence*). The relationship represented by our fifth and last symbol is biconditional. The expression $p \longleftrightarrow q$ is read "p if, and only if, q." "Snow is black if, and only if, all men are mortal." Here we have what is known in mathematics as "a necessary and sufficient condition" and we can call the logical concept "if, and only if."

As we read over these definitions and the examples given for the relation between p and q as expressed by each of the symbols, we are naturally troubled by the fact that they do not seem, according to our understanding of the word, very *logical*. Snow is not black and what does all men being mortal have to do with snow, anyway? Surely the sentential calculus does not concern itself with such inanities!

Let us consider these objections in order.

First: the appropriateness of the examples. In the sentential calculus, p and q, or whatever other variables we use, stand for mathematical propositions. These propositions may be true (*All men are mortal*), or they may be false (*Snow is black*). We are not concerned with their truth or falsity except as it affects the soundness (or logical truth) of the reasoning which follows from them. This important fact is emphasized when the propositions are selected outside the subject matter of mathematics.

Let us take, as an example, one of the simplest and most obvious of the laws of the sentential calculus—the Law of Identity.

$$p \rightarrow p \quad \text{or} \quad \text{If } p, \text{ then } p.$$

If we substitute for the variable p, the "false" statement "Snow is black," we then get the *logically true* statement: "If snow is black, then snow is black." This is just as sound reasoning as that represented by "If all men are mortal, then all men are mortal." A logically false, or unsound, statement is equally false whether p and q are themselves true or false. If, instead of $p \rightarrow p$, we take

$$p \longleftrightarrow \sim p$$

we find that it is as logically false when p stands for "Snow is black," which is false, as it is when p stands for "All men are mortal," which is true. "Snow is black if, and only if, snow is not black." "All men are mortal if, and only if, all men are not mortal." Both are unsound reasonings.

The first hurdle we must overcome is this: We must understand that the truth or falsity of p and q does not directly determine the truth or falsity of the reasoning which is based upon them. The second hurdle is much more difficult.

We were originally bothered by the statement that snow is black, but we were much more bothered about the fact that a statement about snow and one about mortality were combined. Snow and mortality, we objected, have nothing to do with each other; it isn't logical to combine them in one statement! We shall not at this point bring up the common poetic symbolism of winter and death, but shall content ourselves with the comment that it is quite difficult to determine with finality whether two ideas have or do not have something to do with each other.

A simple example will serve. A says, "B attended the University of X and he is a Communist." Obviously, A considers these two ideas related. In the newspaper he has noted that a couple of people recently revealed as Communists attended the University of X. Some of those

262

crackpot professors, he thinks, must be turning the kids into Commies! He connects the two facts that B is a Communist and attended the University of X—connects them both in his mind and in his sentence. C, who is an alumnus of the University of X, objects. There is no connection between the two facts. They do not belong in the same sentence. It is not logical to put them together! Who is right?

If such are the difficulties of determining "relationship" in everyday life, how can we hope to make such a concept precise? The logician answers this question and solves this problem by announcing in a firm voice that, for his purposes, it doesn't matter whether two sentences joined by a symbol of the sentential calculus are, or are not, related. A conjunction $p \wedge q$ will be true if p and q are both true. "Snow is white and all men are mortal" is a completely acceptable sentence from the point of view of the logician. Before we object (we who use "and" too and feel that we have as much right as he to express our opinion), let us remember that the logician does not even suggest that we be governed by the same rule when we use "and." He only says that, for the purpose of developing a calculus with which he can test the logical soundness of mathematical propositions, he must have an unambiguous rule for joining two sentences with "and." As an alumnus of the University of X he would probably argue heatedly with the rest of us about the "logic" of the compound sentence which joins "B attended the University of X" and "B is a Communist." As a logician, examining the proposition, he will say that A's statement is logically sound if it is true that B attended the University of X and if it is also true that B is a Communist.

In the sentential calculus we are concerned with the truth of certain combinations of sentences effected by "not," "and," "or," "if, then," and "if, and only if." We ignore com-

263

pletely any questions of subjective relationship, like *Should these two ideas be put together in the same sentence?* Instead, we concentrate upon the objective relationship. When we put $\sim$ in front of p, the resulting sentence $\sim p$ can be true only if p is false. When we put $\wedge$ between p and q, the resulting sentence $p \wedge q$ can be true only when p and q are both true. Once we accept the idea that p and q do not have to "belong" in the same sentence, we have no objection to these rules.

There are similar arbitrary rules for determining the truth of combinations made with the other symbols. These five symbols, and the logical concepts which they express, are no longer common expressions of everyday discourse, but the technical terms of the sentential calculus:

Not. The sentence $\sim p$ is true only when p is false.

And. The sentence $p \wedge q$ is true only when p and q are both true.

Or. The sentence $p \vee q$ is true if either p or q is true.

If, then. The sentence $p \to q$ is always true except when q is false and p is true.

If, and only if. The sentence $p \longleftrightarrow q$ is true only when p and q are both true or both false.

These definitions of the conditions under which $\sim p$, $p \wedge q$, $p \vee q$, $p \to q$ and $p \longleftrightarrow q$ are true certainly ignore our everyday insistence upon a relationship between two sentences which are joined as one. To determine the logical truth of a combination, we do not even have to know what sentences the variables p and q represent. Given that p is true and q is false, we know that

$\sim p$ is false while $\sim q$ is true;

$p \wedge q$ is false, but $p \vee q$ is true;

$p \to q$ and $p \longleftrightarrow q$ are both false.

To test his understanding of these rules, the reader might like to mark the sentences below "true" or "false" from the point of view of a logician.

p = Snow is white.
q = All men are mortal.

1. Snow is not white.   *T*   *F*
   $\sim p$
2. Snow is white and all men are mortal. *T*   *F*
   $p \wedge q$
3. Snow is white or all men are mortal. *T*   *F*
   $p \vee q$
4. If snow is white, then all men are *T*   *F*
   mortal.
   $p \rightarrow q$
5. Snow is white if, and only if, all men *T*   *F*
   are mortal.
   $p \longleftrightarrow q$

p = $2 + 2 = 5$
q = $2 \times 3 = 4$

6. $2 + 2 \neq 5.$ *T*   *F*
7. $2 + 2 = 5$ and $2 \times 3 = 4.$ *T*   *F*
8. $2 + 2 = 5$ or $2 \times 3 = 4.$ *T*   *F*
9. If $2 + 2 = 5$, then $2 \times 3 = 4.$ *T*   *F*
10. $2 + 2 = 5$ if, and only if, $2 \times 3 = 4.$ *T*   *F*

True Sentences: 2, 3, 4, 5, 6, 9, 10.
Note that in Sentences 1-5, p and q were both true while in 6-10 they were both false.

For every p and q, we have four possible situations: the sentences which p and q represent can be both true,

265

both false, p can be true and q false, or q can be true and p false. As we saw from our examples above, each of these situations may result in a change in the truth or falsity of the combination of p and q effected by a logical symbol. These various possibilities can be stated most simply in the form of a table. In the first column we list by T and F the different possible situations in regard to the truth or falsity of the sentences represented by p and q. The remaining columns are allotted to the different logical relationships; for each we indicate the truth or falsity of that particular combination under the situation regarding p and q as indicated in the first column.

Since the table for the combination effected by $\sim$, or "not," is much simpler than that for the others, we shall give it separately and first.

p	$\sim p$
T	F
F	T

In the following table for the four other combinations, the T's and F's in the first and fourth rows across give us the correct answers to sentences 2-5 and 7-10 in our test on p. 265.

p	q	$p \wedge q$	$p \vee q$	$p \rightarrow q$	$p \longleftrightarrow q$
T	T	T	T	T	T
T	F	F	T	F	F
F	T	F	T	T	F
F	F	F	F	T	T

It is important for us to note that in each of the columns representing a combination of p and q by one of our symbols, we have at least one F. This means that for at least one of the possible situations regarding the truth or

266

falsity of p and q their combination into one statement cannot be regarded as a "true" or logically sound statement. When, however, we construct the same type of table for what in the sentential calculus is called the Law of Identity, or $p \rightarrow p$, which we mentioned earlier, we find that regardless of the truth or falsity of p the combination $p \rightarrow p$ is always true.

p	$p \rightarrow p$
T	T
F	T

Since $p \rightarrow p$ is always true, we say that it is a true sentence. All such true sentences are laws of the sentential calculus and, as we have seen, this is the Law of Identity. We cannot be blamed if we are not too impressed with the Law of Identity. If p, then p. So p implies p. We are reminded of the word *tautology*. Our Law of Identity is certainly tautological. Webster says, "With needless repetition, as *visible to the eye, audible to the ear.*" Logicians say, "A tautology is a true sentence, or law, of the sentential calculus."

The most profound mathematical truths are as tautological as $p \rightarrow p$, but because of their complexity we do not so immediately or intuitively recognize the quality in them. This is where the sentential calculus is indispensable. By means of its so-called truth tables there is a general method for determining whether any statement (no matter how extensive or complicated) is a tautology—in other words, a logically true statement.

The table which we constructed for $p \rightarrow p$ is the simplest possible example of a truth table. As our sentences to be tested increase in the number of their relationships and the number of variables involved, so do their truth

267

tables increase in complexity. Let us take a statement a little more complicated than the Law of Identity and by constructing its truth table determine whether it, too, is a law of the sentential calculus:

$$(\sim p \to p) \to p, \text{ or "If not } p \text{ implies } p, \text{ then } p\text{."}$$

The method which we follow to test this statement is the same one which we will follow for more complicated statements. We take the sentence, beginning most simply, combination by combination.

1. Against the possible truth or falsity of p, we test $\sim p$ in column 2.

2. Against the respective possibilities for p and $\sim p$, we test the combination $\sim p \to p$ in column 3.

3. Against the respective possibilities for $\sim p \to p$ in column 3 and p in column 1, we test the entire sentence $(\sim p \to p) \to p$ in column 4.

p	$\sim p$	$\sim p \to p$	$(\sim p \to p) \to p$
T	F	T	T
F	T	F	T

Since, whether p is true or is false, the statement $(\sim p \to p) \to p$ is always true (as we see in column 4), we know that it is a law of the sentential calculus, or a tautology.

Since any sentence of the calculus can be tested for truth or falsity by means of truth tables, the sentential calculus is one of the few branches of mathematics which has a general method for solving all *its* problems. This almost unique quality of the sentential calculus is extremely significant when we realize that almost all scientific reasoning is based either directly or indirectly upon its laws. We are then, in the words of Tarski, able to dissect even the most complicated mental processes by "such simple activities as attentive observation of statements previously accepted as true, the perception of structural, purely external connections among these statements, and the execution of mechanical transformations as prescribed by the rules of inference. It is obvious that, in view of such a procedure, the possibility of committing mistakes in a proof is reduced to a minimum."

This achievement of the sentential calculus is all the more impressive when we consider the simplicity of the tools with which it works—half a dozen concepts expressed by some of the simplest words in the language. It is an achievement that Euclid would have appreciated.

FOR THE READER

Following the method of the truth tables which we have detailed in this chapter, the reader can now determine for each of the following two sentences whether it is a true sentence in the sentential calculus. (One is and one isn't.)

$$(p \to q) \longleftrightarrow (q \to p)$$
$$[(p \to q) \to p] \to p$$

p	q	$p \to q$	$q \to p$	$(p \to q) \longleftrightarrow (q \to p)$
T	T	T	T	
T	F	F	T	
F	T	T	F	
F	F	T	T	

p	q	$p \to q$	$(p \to q) \to p$	$[(p \to q) \to p] \to p$
T	T	T	T	
T	F			
F	T			
F	F			

On filling in the blanks, we see that the second sentence (but not the first) is a tautology.

The reader can now by the same method construct a truth table for a fairly complicated statement:

If p implies q and q implies r, then p implies r

When we transcribe this sentence into the symbolism of the sentential calculus, we get the statement below.

$$[(p \to q) \wedge (q \to r)] \to (p \to r)$$

To construct a truth table for this sentence, we must first list the possibilities in regard to the truth or falsity of the three variables, p, q and r. We then check off against these possibilities the truth or falsity of the logical combinations of the variables in the following somewhat nested order:

$$p \to q$$
$$q \to r$$
$$(p \to q) \wedge (q \to r)$$
$$p \to r$$
$$[(p \to q) \wedge (q \to r)] \to (q \to r)$$

We leave it to the reader to determine whether this is a law of the sentential calculus.

The reader should come out with a T in every space in the last column. If he substitutes for p, "Snow is black," and for q, "All men are mortal," and for r, "Columbus discovered America last year," he will find that the statement is indeed a tautology and therefore a law of the sentential calculus.

270

18

**Mathematics,
the
Inexhaustible**

AT THE MID-POINT OF THE TWENTIETH century, more than two thousand years after Euclid compiled his *Elements,* the axiomatic method—the method which is synonymous with the *Elements* of Euclid—was the subject of an international symposium of mathematicians and scientists, the primary purpose of which was to determine the extent to which this classic method of mathematics could and should be further utilized by the physical sciences. The discussions, the concepts, and the vocabulary were a long *long* way from Euclid—and they were perhaps farthest in the discussions of elementary geometry!

"What is elementary geometry?" asked Alfred Tarski, the famed logician, and answered as follows:*

"We regard as elementary that part of Euclidean geometry which can be formulated and established without the help of any set-theoretical devices."

Tarski then continued with a more precise statement of his view of elementary plane geometry (or E_2) as formulated in the terms of first-order predicate calculus, which is printed in full on page 272.

---

* The following quotation and the axioms of elementary geometry appear in *The Axiomatic Method,* Leon Henkin, Patrick Suppes and Alfred Tarski, editors, North-Holland Publishing Company, Amsterdam, The Netherlands, 1959.

All the variables $x, y, z, \ldots$ occurring in this theory are assumed to range over elements of a fixed set; the elements are referred to as points, and the set as the space. The logical constants of the theory are (i) the sentential connectives—the negation symbol $\sim$, the implication symbol $\rightarrow$, the disjunction symbol $\vee$, and the conjunction symbol $\wedge$; (ii) the quantifiers—the universal quantifier $\wedge$ and the existential quantifier $\vee$; * and (iii) two special binary predicates—the identity symbol $=$ and the diversity symbol $\neq$. As non-logical constants (primitive symbols of the theory) we could choose any predicates denoting certain relations among points in terms of which all geometrical notions are known to be definable. Actually we pick two predicates for this purpose: the ternary predicate β used to denote the betweenness relation and the quaternary predicate δ used to denote the equidistance relation; the formula $\beta(xyz)$ is read *y lies between x and z,* (the case when y coincides with x and z not being excluded), while $\delta(xyzu)$ is read *x is distant from y as z is from u.* †

Strange though the language of Tarski's twentieth-century geometry might seem to Euclid, it would not be so far removed from the Greek as the simple statement of Tarski's Theorem 3:

THEOREM 3: *The theory E_2 is decidable.*

* The universal quantifier $\wedge$ stands for "for every" and the existential quantifier $\vee$, for "there exists."

† Using the vocabulary which is given here, the reader may enjoy translating *into words* Tarski's axioms for elementary geometry.

This five-word theorem embodies that aspect of modern mathematics which is undoubtedly farthest from Euclid—what has been called "the most significant truth" of the twentieth century!

In this chapter we shall try to give the reader a glimpse of this truth by clarifying the meaning of that deceptively simple word, *decidable,* in the statement of the theorem above. In general terms, the statement that the theory E_2 is decidable means that for elementary plane geometry, as formulated by Tarski in the paragraphs above, there exists *a method* for solving all possible problems.

What do we mean by a method of solving an infinite class of problems, such as *all* the problems of elementary plane geometry? This is a question to which Euclid's successors of the twentieth century have devoted considerable thought, and the answer they have at length come to is among the most significant in the history of mathematics.

Curiously, their interest in what they meant by a method developed from the consideration, suggested for the first time by Gödel, that for some classes of mathematical problems there might be no method. This is understandable. If someone comes to us and says, "I have a method of doing so and so," we do not stop him with, "See here. Just what do you mean by a method?" Instead we say, "What is it?" It is only when he comes and says, "There is no method of doing so and so," that we stop him with, "Just what do you mean when you say there is no method?"

This is essentially the situation that occurred in mathematics in 1931. In that year Gödel, as we have told in Chapter 16, published a paper "On Formally Undecidable Propositions of *Principia Mathematica* and Other Related Systems." This was one of the great turning points

273

in mathematical thought. Although the paper was concerned primarily with demolishing the idea that the absolute consistency of a mathematical system could be established within that system, implicit in it was the idea that for certain classes of problems (such as those encountered in number theory), there can be no general method of solving all of the problems in the class.

This truly monumental result started other mathematicians thinking for the first time upon the subject of methods in general. What did they mean by a method? Working more or less independently here and abroad, several of them formulated definitions of a method. Most definitions were extremely technical (one of the most important depending upon the idea of recursive functions); but there was one among them the mere name of which evokes a refreshingly non-mathematical image. This particular definition of a method was put forth by A. M. Turing (1912-1954) and is called a *Turing machine*.

Since the mechanical way of thinking was almost as natural to Turing as the mathematical, it is not surprising that when he set out to define a method, he thought of it as something which could be performed by a machine. Said Turing: If a machine could be conceived of as solving an arbitrarily chosen problem of an infinite class, then indeed we have a general method for that class of problems. When we say there is no method of solving an infinite class of problems, we mean that it is impossible to conceive of such a machine.

With a method, according to this definition, a machine could be given a set of specific instructions which it would follow for a finite length of time, depending upon the particular problem of the class that it was given; and eventually—perhaps eons from now—it would turn out an answer, the right answer, to that problem. Instructions

274

for the machine would have to be absolutely determined in advance: do some specific thing until some other specific thing happens and then do some specific other thing. The machine could ask no questions, exercise no judgments, make no innovations. Each problem would have to go in, and come out, with every step toward its solution automatically decided by the method alone. Otherwise, no method.

Such a machine as Turing conceived is not even meant to be constructible. Conceptually, it is very like one of the great electronic computing machines which are in existence at the present time. In many ways it is conceived of as being less efficient than they, for its aim is not efficiency but simplicity. In other ways it is (quite literally) infinitely more efficient. It is in the nature of the infinite classes of problems with which we are dealing that, while a computer may be in a sense "close" to a Turing machine, it can never—in spite of all possible improvements in its efficiency—be any "closer." This becomes clear when we consider a specific and infinite class of problems for which a general method has been known since before the time of Euclid. *Is a given number* n *a prime?* Theoretically, we can solve this problem for any n by attempting to divide it by every prime which is smaller than (or equal to) $\sqrt{n}$; if none of these divides it, then n is a prime. Practically, though, we find very soon that n is too large for us to test by this method. Although mathematicians have devoted years to testing the primality of certain interesting numbers, life is literally too short to accomplish this, and they must yield to the electronic computing machines. But very soon n is too large for the machines. The largest number which has been tested and found prime is $2^{9941} - 1$. By everyday standards $2^{9941} - 1$ is quite a large number, being some

275

3000 digits in length; yet among the primes it is a relatively small one. Since there are only a finite number of primes which are smaller than $2^{9941} - 1$ but an infinite number of primes which are larger, "almost all" primes are larger than the largest known prime. Obviously, an actual machine, because of the limits of time and storage, can never solve all or even certain specific problems of an infinite class. A Turing machine, being purely conceptual, has no such limits because it is conceived of as having an arbitrarily large amount of time and an arbitrarily large memory or storage—as large as it needs for any given problem in a class. Only for this reason is it unconstructible.

The mathematical point to the Turing machine is not whether there could or could not be such a machine. A Turing machine is simply a set of specifications, not for a machine, but for a method of solving an infinite class of mathematical problems. The limits imposed by the concept of the machine upon a method are as follows:

The machine is allowed an arbitrarily large amount of time in which to solve a problem and an arbitrarily large amount of paper on which to do the work. A roll of tape keeps moving through it. This tape consists of a series of positions of rest which can be visualized simply as squares. At any particular instant only one of these squares is being scanned by the machine. How the machine reacts is determined by (1) the contents of the square and (2) the internal state of the machine. The square contains one of a finite number of symbols and the machine is in one of a finite number of internal states. On the basis of these two factors, in the time interval allowed, the machine can change the contents of the square, change its position by no more than one square and/or change its internal state. It can have no choice, in the usual sense; what it *does* is absolutely determined by the method. Also included is a way of feeding problems to the machine and of recognizing when the machine has finished a problem.

Such is the conceptual blueprint for a Turing machine. If what we call a method for solving an infinite class of problems (like determining whether or not n is prime) can be used within these limitations to solve any arbitrary problem of the class, then we have a method. When we say that there is no method for solving such an infinite class of problems, we mean that the class includes problems which cannot by their nature be solved by such a machine.

By a method we mean a machine.

Perhaps this does not sound like what we usually consider a precise definition; yet when we begin to apply it, we find that it does define what we mean by a method, and very precisely. The method for determining whether or not a given n is prime is a method in this sense; for, as we have seen, determining primality by machine is common practice and limited only by physical considerations of time and storage.

In the preceding chapter we described the method of truth tables by which it is possible to determine whether any sentence of the sentential calculus is a true sentence and, therefore, a law of the calculus. It is easily seen that this, too, is a general method according to our definition of a method as a machine. We can conceive of a Turing machine which, using the method of truth tables, could solve any of the problems of the sentential calculus no matter how long and complicated the sentences involved might be. Since all of its problems are solvable by such a general method, we call the sentential calculus a *decidable theory*.

Tarski's Theorem 3, which we gave at the beginning of

this chapter, tells us merely that elementary geometry is also a decidable theory. The more limited a class of problems (even though the class is infinite), the more likely it is that there exists a general method of solving all the problems in the class. The sentential calculus is the most fundamental and elementary theory of logic and is, as we have seen, a decidable theory. First-order predicate calculus, a step above it in complexity and importance, is an *undecidable theory*. The theory of numbers—defined as all those problems which can be expressed in terms of the integers, the basic concepts of logic, and multiplication and addition—is an undecidable theory, as Kurt Gödel showed in 1931. When we take a more limited class of number problems, like those of elementary arithmetic, we find that we have a decidable theory.

Sometimes, however, when we enlarge our definition, we get a decidable theory. When, as in the case of the problems of elementary algebra, we define our class in the same terms by which we define the problems of number theory except for the fact that we substitute the real numbers for the integers, we find that we have a decidable theory. Interestingly, Tarski's proof that elementary geometry is a decidable theory follows from the proof (also his) that elementary algebra is a decidable theory, elementary geometry and elementary algebra being both concrete representations of the same abstract theory.

In the last quarter of a century, as a result of the precise defining of method by Turing and others, modern mathematicians have been able to till a field which was undreamed of by their predecessors: the determination of undecidable theories, those classes of mathematical problems for which there can be no general method. Just how undreamed-of this field is can best be illustrated by a famous problem proposed at the turn of the century by

David Hilbert. As the leading mathematician of the day, he gave to his colleagues a list of problems which he felt needed to be solved. One of these was to determine a general method of solution for all indeterminate, or Diophantine, equations. These, a sub-class of the problems of number theory, take their name from Diophantus of Alexandria, who had a fondness for them. These are problems in two or more unknowns for which integer solutions are required. A simple example is $x^2 - y^3 = 17$, which is one of an infinite class of problems represented by the equation $x^2 - y^3 = n$, in turn a sub-class of the class of all Diophantine problems.

When Hilbert, in 1900, proposed to his colleagues that they attempt to determine a general method for solving all Diophantine problems, he—and his colleagues, as well—assumed that such a general method existed. Today—so great have been the recent developments in meta-mathematics *—it is generally considered probable (although such has not yet been proved) that there can be no general method for solving all Diophantine problems: that it is an undecidable theory. Even its relatively small sub-class, mentioned above, presents difficulties. It is not known whether there is a general method for solving the class of problems $x^2 - y^3 = n$. Such problems have only a finite number of solutions. This has been proved. For instance, the specific problem $x^2 - y^3 = 17$, already mentioned, has the following solutions when x is positive:

x	3,	4,	5,	9,	23,	282,	375,	378,661
y	—2,	—1,	2,	4,	8,	43,	52,	5,234

These solutions were obtained by a "method" which works in a great many cases—in fact, has never failed to work in any case; yet it has never been shown—in the sense of a

* The study of the structure of mathematics.

279

method such as that which can be performed by a machine —that it will work in all cases.

To show that the class of problems $x^2 - y^3 = n$ is decidable, someone must prove that this or some other method is a truly general method which could be used by a machine to solve any arbitrary problem of the class. To show that the class is undecidable, someone must establish that in it there exist problems, $x^2 - y^3 = n$, which by their nature cannot be solved by any general method. It is quite likely that this particular class of problems is decidable and that the known method is truly general. If, however, someone were to prove tomorrow that the class is undecidable, the result would have great significance: for, by establishing the undecidability of a sub-class of Diophantine problems, it would at the same time establish the undecidability of the class of all Diophantine problems.

In such a way the determination of undecidable theories—sub-classes in themselves of all mathematics—establishes, as well, a fact of overwhelming significance: that mathematics itself is undecidable. The answer to the question

Can there be a general method for solving all mathematical problems?

is *no!*

Perhaps, in a world of unsolved and apparently unsolvable problems, we would have thought that the desirable answer to this question, from any point of view, would be *yes*. But from the point of view of mathematicians a *yes* would have been far less satisfying than a *no* is. Now it is established—with all the certainty of logical proof—that machines can never, even in theory, replace mathematicians.

280

The language of twentieth-century elementary geometry, a curious combination of logic and letters, is a long way from Euclid. Decision theory was undreamed of in his mathematics; yet the conclusion to which mathematics has come as a result of Gödel's paper would be as satisfying to Euclid as to any mathematician of the twentieth century:

Not only are the problems of mathematics infinite and hence inexhaustible, but mathematics itself is inexhaustible.

FOR THE READER

We have come a long way from Euclid, and perhaps how very far we have traveled is shown most vividly by a comparison of Euclid's axioms, which appear on page 27, and those of Tarski's E_2, which are printed in full below:

A1   [Identity Axiom for Betweenness].
       $\wedge\, xy[\beta(xyx) \to (x = y)]$

A2   [Transitivity Axiom for Betweenness].
       $\wedge\, xyzu[\beta(xyu) \wedge \beta(yzu) \to \beta(xyz)]$

A3   [Connectivity Axiom for Betweenness].
       $\wedge\, xyzu[\beta(xyz) \wedge \beta(xyu) \wedge (x \neq y) \to$
             $\beta(xzu) \vee \beta(xuz)]$

A4   [Reflexivity Axiom for Equidistance].
       $\wedge\, xy[\delta(xyyx)]$

A5   [Identity Axiom for Equidistance].
       $\wedge\, xyz[\delta(xyzz) \to (x = y)]$

A6   [Transitivity Axiom for Equidistance].
       $\wedge\, xyzuvw[\delta(xyzu) \wedge \delta(xyvw) \to \delta(zuvw)]$

A7   [Pasch's Axiom].
       $\wedge\, txyzu \vee v[\beta(xtu) \wedge \beta(yuz) \to \beta(xvy) \wedge$
                $\beta(ztv)]$

A8   [Euclid's Axiom].
$\wedge\ txyzu \vee vw[\beta(xut) \wedge \beta(yuz) \wedge (x \neq u) \rightarrow$
$\beta(xzv) \wedge \beta(xyw) \wedge \beta(vtw)]$

A9   [Five Segment Axiom].
$\wedge\ xx'yy'zz'uu'[\delta(xyx'y') \wedge \delta(yzy'z') \wedge \delta(xux'u')$
$\wedge\ \delta(yuy'u') \wedge \beta(xyz) \wedge \beta(x'y'z')$
$\wedge\ (x \neq y) \rightarrow \delta(zuz'u')]$

A10   [Axiom of Segment Construction].
$\wedge\ xyuv \vee z[\beta(xyz) \wedge \delta(yzuv)]$

A11   [Lower Dimension Axiom].
$\vee\ xyz[\sim\beta(xyz) \wedge \sim\beta(yzx) \wedge \sim\beta(zxy)]$

A12   [Upper Dimension Axiom].
$\wedge\ xyzuv[\delta(xuxv) \wedge \delta(yuyv) \wedge \delta(zuzv) \wedge$
$(u \neq v) \rightarrow \beta(xyz) \vee \beta(yzx) \vee$
$\beta(zxy)]$

A13   [Elementary Continuity Axioms].
All sentences of the form
$\wedge\ vw \ldots \{ \vee z \wedge xy[\Phi \wedge \Psi \rightarrow \beta(zxy)] \rightarrow$
$\vee u \wedge xy[\Phi \wedge \Psi \rightarrow \beta(xuy)]$

where Φ stands for any formula in which the
variables $x, v, w, \ldots$, but neither y nor z nor u,
occur free, and similarly for Ψ, with x and y
interchanged.

If the reader will take in hand pencil and paper and
the vocabulary for E_2 from page 272, he will be surprised to find how easily he can translate some of these
axioms into statements which will be meaningful for him.
(The first few particularly!)

But it's a *long* way!

Index

absolute value, 143, 215
addition (*see also* arithmetic, operations of), 99-101, 138-140, 193-196, 278
aleph-one, 225
aleph-zero (*see also* infinite, theory of), 211, 217, 225
Alexandria, 17, 18, 26, 279
algebra, 67-68, 83, 106, 137, 169, 171, 177
  elementary, 270
  Fundamental Theorem of, 112, 215
  of *n* variables (*see also* geometry, *n*-dimensional), 173
algebraic numbers (*see also* complex numbers; transcendental numbers), 42, 214-217
algebraic processes as geometrical constructions (*see also* construction problems), 138-148
analysis, 43-44, 115, 177
analysis situs (*see also* topology), 191
analytic geometry, 7, 67-77, 82-83, 115-116, 130, 137, 141, 168-169
"and" (*see also* calculus, sentential), 258, 260, 262-264, 266
angle, 149-150, 158, 178, 190
  right, 23, 26, 177
  trisection of (*see also* construction problems), 133-134, 135, 142-145, 146-147, 225
Apollo, 61, 75
Apollonius, 18, 63, 66, 67, 68, 115, 119

283

284

291

truth, geometrical, 148-149, 152, 161

truth or falsity of sentences, 261-262

truth tables (*see also* calculus, sentential), 277

Turing, A. M., 274, 275, 278

Turing machine, 274-277

Turnbull, H. W., 17

Two Square Theorem, 37-38

unity, roots of, 201

variable, 67

Vatican, 19

volume, 226, 243

whole numbers (*see also* natural numbers), 30, 48, 109

Yeats, William Butler, 17

zero, 102-103, 104, 105, 106, 204, 205